Vincenzo Pallotta, Alessandro Soro, and Eloisa Vargiu (Eds.)

Advances in Distributed Agent-Based Retrieval Tools

Studies in Computational Intelligence, Volume 361

Editor-in-Chief

Prof. Janusz Kacprzyk
Systems Research Institute
Polish Academy of Sciences
ul. Newelska 6
01-447 Warsaw
Poland
E-mail: kacprzyk@ibspan.waw.pl

Further volumes of this series can be found on our homepage: springer.com

Vol. 342. Federico Montesino Pouzols, Diego R. Lopez, and Angel Barriga Barros
Mining and Control of Network Traffic by Computational Intelligence, 2011
ISBN 978-3-642-18083-5

Vol. 343. Kurosh Madani, António Dourado Correia, Agostinho Rosa, and Joaquim Filipe (Eds.)
Computational Intelligence, 2011
ISBN 978-3-642-20205-6

Vol. 344. Atilla Elçi, Mamadou Tadiou Koné, and Mehmet A. Orgun (Eds.)
Semantic Agent Systems, 2011
ISBN 978-3-642-18307-2

Vol. 345. Shi Yu, Léon-Charles Tranchevent, Bart De Moor, and Yves Moreau
Kernel-based Data Fusion for Machine Learning, 2011
ISBN 978-3-642-19405-4

Vol. 346. Weisi Lin, Dacheng Tao, Janusz Kacprzyk, Zhu Li, Ebroul Izquierdo, and Haohong Wang (Eds.)
Multimedia Analysis, Processing and Communications, 2011
ISBN 978-3-642-19550-1

Vol. 347. Sven Helmer, Alexandra Poulovassilis, and Fatos Xhafa
Reasoning in Event-Based Distributed Systems, 2011
ISBN 978-3-642-19723-9

Vol. 348. Beniamino Murgante, Giuseppe Borruso, and Alessandra Lapucci (Eds.)
Geocomputation, Sustainability and Environmental Planning, 2011
ISBN 978-3-642-19732-1

Vol. 349. Vitor R. Carvalho
Modeling Intention in Email, 2011
ISBN 978-3-642-19955-4

Vol. 350. Thanasis Daradoumis, Santi Caballé, Angel A. Juan, and Fatos Xhafa (Eds.)
Technology-Enhanced Systems and Tools for Collaborative Learning Scaffolding, 2011
ISBN 978-3-642-19813-7

Vol. 351. Ngoc Thanh Nguyen, Bogdan Trawiński, and Jason J. Jung (Eds.)
New Challenges for Intelligent Information and Database Systems, 2011
ISBN 978-3-642-19952-3

Vol. 352. Nik Bessis and Fatos Xhafa (Eds.)
Next Generation Data Technologies for Collective Computational Intelligence, 2011
ISBN 978-3-642-20343-5

Vol. 353. Igor Aizenberg
Complex-Valued Neural Networks with Multi-Valued Neurons, 2011
ISBN 978-3-642-20352-7

Vol. 354. Ljupco Kocarev and Shiguo Lian (Eds.)
Chaos-Based Cryptography, 2011
ISBN 978-3-642-20541-5

Vol. 355. Yan Meng and Yaochu Jin (Eds.)
Bio-Inspired Self-Organizing Robotic Systems, 2011
ISBN 978-3-642-20759-4

Vol. 356. Slawomir Koziel and Xin-She Yang (Eds.)
Computational Optimization, Methods and Algorithms, 2011
ISBN 978-3-642-20858-4

Vol. 357. Nadia Nedjah, Leandro Santos Coelho, Viviana Cocco Mariani, and Luiza de Macedo Mourelle (Eds.)
Innovative Computing Methods and Their Applications to Engineering Problems, 2011
ISBN 978-3-642-20957-4

Vol. 358. Norbert Jankowski, Włodzisław Duch, and Krzysztof Grąbczewski (Eds.)
Meta-Learning in Computational Intelligence, 2011
ISBN 978-3-642-20979-6

Vol. 359. Xin-She Yang and Slawomir Koziel (Eds.)
Computational Optimization and Applications in Engineering and Industry, 2011
ISBN 978-3-642-20985-7

Vol. 360. Mikhail Moshkov and Beata Zielosko
Combinatorial Machine Learning, 2011
ISBN 978-3-642-20994-9

Vol. 361. Vincenzo Pallotta, Alessandro Soro, and Eloisa Vargiu (Eds.)
Advances in Distributed Agent-Based Retrieval Tools, 2011
ISBN 978-3-642-21383-0

Vincenzo Pallotta, Alessandro Soro, and
Eloisa Vargiu (Eds.)

Advances in Distributed
Agent-Based Retrieval Tools

Editors

Vincenzo Pallotta
InterAnalytics
Rue des Savoises, 19
1205 Geneva, Switzerland
E-mail: vincenzo.pallotta@interanalytics.ch

Eloisa Vargiu
Department of Electrical and
Electronic Engineering
University of Cagliari
Piazza d'Armi
09123 Cagliari – Italy
E-mail: vargiu@diee.unica.it

Alessandro Soro
CRS4, Center of Advanced Studies Research
and Development in Sardinia
Parco Scientifico della Sardegna, Ed. 1
09010 Loc. Piscinamanna, Pula, (CA) – Italy
E-mail: asoro@crs4.it

ISBN 978-3-662-50675-2

e-ISBN 978-3-642-21384-7

DOI 10.1007/978-3-642-21384-7

Studies in Computational Intelligence

ISSN 1860-949X

© 2011 Springer-Verlag Berlin Heidelberg

Typeset & Cover Design: Scientific Publishing Services Pvt. Ltd., Chennai, India.

Printed on acid-free paper

9 8 7 6 5 4 3 2 1

springer.com

Preface

Nowadays, Internet and the Web are not only vehicles of unstructured and heterogeneous contents. In fact, they are rapidly morphing into a platform for interaction and collaboration. In this scenario, the goal of next generation information retrieval tools will be to support semantics, personalization, context awareness and seamless access to highly variable data and messages coming from document repositories, social media and networks, and ubiquitous sensors and devices.

The DART workshop series on Distributed Agent-Based Retrieval Tools was launched in 2006 and reached its 4th edition in 2010. It was originally aimed at putting together practitioners and researchers working on novel retrieval tools for distributed systems and environments. The main goal was to contribute to the discussion about advances in pervasive and intelligent access to Web services and distributed information systems. Over time, the aim of the workshop has been extended to the role of semantics and distributed information retrieval to support content-based multimedia indexing and search.

This book collects revised and extended versions of articles submitted and accepted for presentation to the DART'10 workshop. The 4th edition of DART was held in Geneva at the Webster University international campus, on June 2010.

Social media and collaboration are the topics of the first two chapters.

In Chapter 1, *Rethinking Search Engines in Social Network Vision*, Angioni et al. focus on re-contextualizing the current search engine technology to social networks. In particular, they study the integration of Semantic Web and Natural Language Processing (NLP) technology in a unifying framework. In so doing, user-centric points of views of socially connected users can be used to generate a more effective notion of relevance.

In Chapter 2, *A Collaborative Web Application for Supporting Researchers in the Task of Generating Protein Datasets*, Armano and Manconi describe a collaborative web application for supporting the generation of protein datasets. In fact, the lack of specific datasets is a major problem for bioinformaticians. ProDaMa-C, the collaborative web application presented in this Chapter, helps bioinformaticians in generating those datasets.

Improving search engines is the main topic of the next three chapters.

In Chapter 3, *RefGen: Identifying Reference Chains to Detect Topics*, Longo and Todirascu propose advanced NLP techniques for improving search engine performance through topic indexing. Their approach for topic detection is based on the identification of references chains.

In Chapter 4, *Synonym Acquisition Across Domains and Languages*, Van der Plas et al. contribute to improve quality of search by proposing an automatic method for acquiring synonyms over different domains and languages. Synonyms can then be used for query expansion for enlarging the search scope by exploiting semantics.

Improving the quality of search is also the goal of Chapter 5, *Linguistically-based Reranking of Googles Snippets with GreG*, in which Delmonte and Tripodi propose a re-ranking approach to standard search engine. The approach is based on the linguistic analysis of unconstrained natural language queries.

The next two chapters are concerned with Sentiment Analysis and Opinion Mining.

In Chapter 6, *Opinion Mining and Sentiment Analysis Need Text Understanding*, Delmonte and Pallotta advocate for the need of full linguistic processing of input in order to achieve accurate and robust sentiment analysis of product reviews.

At the opposite extreme, in Chapter 7, *Sentiment Analysis of French Movie Reviews*, Ghorbel and Jacot describe a Machine-Learning approach focused on classifying the polarity (positive, negative) of conveyed opinions. The authors present a supervised classification of French movie reviews based on some shallow linguistic features such as part-of-speech tagging and word semantic orientation.

Distributed information retrieval is tackled in Chapter 8 and 9.

In Chapter 8, *Query Building in a Distributed Semantic Indexing System*, Moulin and Lai propose a query expansion mechanism to deal with terminology variation of semantic description of shared resources in peer-to-peer networks.

In Chapter 9, *Building Distributed and Pervasive Information Management Systems with HDS*, Bergenti and Poggi present the Heterogeneous Distributed System (HDS) framework built on top of JADE, one of the most established agent framework. The chapter also present the RAIS (Remote Assistant for Information Sharing) application built with HDS whose goal is to support the sharing of information among a community of connected users.

Pervasive intelligence is the main topic of the last two chapters.

In Chapter 10, *Sensor Mining for User Behavior Profiling in Intelligent Environments*, Augello et al. consider the problem of mining information from sensor networks for user profiling purposes. They report the results of an experiment carried out in office aimed at building a probabilistic model that could be used to predict users behavior and optimize the deployment of building resources.

Pervasive intelligence is also used to build a context-aware recommendation system for exhibition tours planning. In Chapter 11, *Motivating Serendipitous Encounters in Museum Recommendations*, Iaquinta et al. evaluate the impact in users satisfaction of introducing novelty and diversity in personalized museum tours.

We would like to thank all the authors for their excellent contributions and the reviewers for their careful revision and suggestions for improving them. We are grateful to the Springer-Verlag Team for their assistance during preparation of the manuscripts.

We are also indebted to all the participants and scientific committee members of the four editions of the DART workshop, for their continuous encouragement, support and suggestions.

March 2011   Vincenzo Pallotta
                                                                         Alessandro Soro
                                                                          Eloisa Vargiu

Contents

Rethinking Search Engines in Social Network Vision......... 1
Manuela Angioni, Emanuela De Vita, Cristian Lai, Ivan Marcialis,
Gavino Paddeu, Franco G. Tuveri

A Collaborative Web Application for Supporting
Researchers in the Task of Generating Protein Datasets 13
Giuliano Armano, Andrea Manconi

RefGen: Identifying Reference Chains to Detect Topics 27
Laurence Longo, Amalia Todiraşcu

Synonym Acquisition across Domains and Languages 41
Lonneke van der Plas, Jörg Tiedemann, Jean-Luc Manguin

Linguistically-Based Reranking of Google's Snippets with
GreG 59
Rodolfo Delmonte, Rocco Tripodi

Opinion Mining and Sentiment Analysis Need Text
Understanding.. 81
Rodolfo Delmonte, Vincenzo Pallotta

Sentiment Analysis of French Movie Reviews 97
Hatem Ghorbel, David Jacot

Query Building in a Distributed Semantic Indexing
System 109
Claude Moulin, Cristian Lai

Building Distributed and Pervasive Information
Management Systems with HDS 129
Federico Bergenti, Agostino Poggi

Sensor Mining for User Behavior Profiling in Intelligent Environments... 143
Agnese Augello, Marco Ortolani, Giuseppe Lo Re, Salvatore Gaglio

Motivating Serendipitous Encounters in Museum Recommendations 159
Leo Iaquinta, Marco de Gemmis, Pasquale Lops, Giovanni Semeraro, Piero Molino

Author Index... 169

Rethinking Search Engines in Social Network Vision

Manuela Angioni, Emanuela De Vita, Cristian Lai, Ivan Marcialis,
Gavino Paddeu, and Franco G. Tuveri

Abstract. In this chapter we illustrate our vision about the evolution of search engines, dealing with some emerging questions related to the social role of the user on the Web and to the actual approach to access the information. In this scenario, is ever more evident the need to redefine the information paradigm bringing the information to the user and not more the user to the information, with search engines able to provide results without direct questions from users, anticipating their needs. A Web in service of the user, automatically informed by the system with suggested resources related with his life style and his common behavior without the need to ask for them. This approach will be applied to a project named A Semantic Search Engine for a Business Network where the development of a business network creates a point of contact between the academic and the research world and the productive one by the introduction of Natural Language Processing, user profiling, automatic information classification according to users' personal schemas, contributing in such a way to redefine the vision of information and delineating processes of Human-Machine Interaction.

1 Introduction

The Web's evolution during the last few years shows that the advantages from the users' point of view are not so macroscopic. It is going more and more toward tools able to follow and assist the user in networking activities through the use of technologies related to natural languages, the classification of the information and the user profile (Marcialis and De Vita, 2008). In this scenario changes carried out by the great innovators in the field of information processing are emerging. Google is still the frontier of search engines, but there are several efforts in order to exceed its capabilities, such as Bing, which provides good results on search suggestions and allows natural language queries. Meta search engines try to reduce the time consumed on online search, allowing users to send queries

Manuela Angioni · Emanuela De Vita · Cristian Lai · Ivan Marcialis
Gavino Paddeu · Franco G. Tuveri
CRS4, Center of Advanced Studies, Research and Development in Sardinia,
Parco Scientifico e Tecnologico, Ed. 1
09010 Pula (CA), Italy
e-mail: {angioni,emy,clai,ciano,paddeu,tuveri}@crs4.it

V. Pallotta, A. Soro, and E. Vargiu (Eds.): Advances in DART, SCI 361, pp. 1–12.
springerlink.com   © Springer-Verlag Berlin Heidelberg 2011

simultaneously on more search engines and aggregating the results, such as Bin-gandGoogle[1] or SortFix[2] , that searches Google, Yahoo and Twitter by means of a drag-and-drop interface that allows the user to describe a detailed and precise query.

In this chapter we illustrate an overview and the ideas behind a project named *A Semantic Search Engine for a Business Network*. It involves the development of a business network able to create a point of contact between the academic and the research world in general and the productive one, with the aim of encouraging the cooperation and the sharing of ideas, of different point of views, information material or needs, and in order to support the productive world and the associated decision-making process. One of the project's objectives is to answer to the questions expressed in the following of the introduction.

Actually the online social networking is becoming more and more popular and several experiments in social network-based Web search have been performed in order to demonstrate the potential for using online social networks to enhance Internet search (Mislove, 2006).

The introduction of queries in natural language is a common element that is already prefiguring the advent of the Web 3.0. An example is Twine (Wissner and Spivack, 2009), able to improve the relevance of results by means of filters that try to reduce the noise due to less relevant answers. Other emerging tools are the computational knowledge engine Wolfram Alpha[3], able to answer queries by means of a vast repository of data organized with the help of sophisticated Natural Language Processing algorithms, or Aardvark (Horowits and Kamvar, 2010) that allows users, experts on certain topics, to answer to queries made by other users in a more efficient mechanism for online search.

Despite information is still the primary element, is ever more evident the need to redefine the information paradigm so that the net and the information become "really" user-centric by an inverse process that brings the information to the user and not more the user to information.

In our opinion, what each user needs is a specific private data strictly related to his point of view, his way to classify and manage the information, his network of contacts in the way everybody choose to live the Web, the net and the knowledge. So, new tools able to reduce or even to eliminate the search phase performed by the user are needed, but certainly commercial search engines, that make profit by the number of access to their pages, are not interested in produce them.

The passage from the unstructured to the structured information through the use of ontologies has not produced the expected innovation in search engines due to the lack of tagged resources.

The rethinking of search engines involves the emerging of some questions about the method of search through repeated queries and their successive refinement. Someone thinks that search engines should be considered "only a primitive form of decision support" (Spivack, 2010). So, the vision of a Web where search

---

[1] http://www.bingandgoogle.com/

[2] http://www.sortfix.com/

[3] http://www.wolframalpha.com/

engines are able to provide results without direct questions from users, anticipating their needs, could be now plausible. A Web of user disposal, automatically informed by the system with suggested resources related with to his life style and his common behaviour without the need to ask for them. Such idea of Web and of search engines above described is applied to a project and will converge in a system able to support and follow users in their activities. In particular the idea behind the project is the realization of a business network able to guarantee the match and the cooperation of academic and research world with the productive one in order to sustain related production and decisional processes.

The reminder of the chapter is organized as follows: section 2 describes a use case. In section 3 are described the aims of the project in a general way; section 4 analyzes the search engine; section 5 shows the set of modules; section 6 explains the idea of business network and finally are described the conclusions.

2 Use Case

Companies and researchers meet on the Web using the usual navigation tools.

The following use case describes the effort to make compatible the needs of researchers and companies, making easier the meeting of their common goals and favouring the transfer between the research knowledge and the companies needs that look for innovative ideas to apply to their own business model.

As depicted in Figure 1, the user browses the Web with Mozilla Firefox, provided with a proper extension. The viewed pages are modified and displayed according to the user profile and his preferences. An intelligent agent observes the user browsing in real time, pointing out all the information that better meet the definition of his profile as interesting.

The user, after the registration, is identified by his account and access to his homepage and to the social network notice board. The registration enables the system to access to the notice board for the analysis of contents. The system analyzes the user contacts list in order to generate a network of users unrelated to the profiles similarity. During his browsing activity the user highlights, annotates, inserts tags and classifies the interesting portions of the visited Web pages.

Moreover, the system shows information related to the current page. Information could include profiles of people with similar interests, companies and projects profiles correlated to the user profile, annotations of parts of pages by similar users and friends and besides, similar pages.

Related to networks of expertise generated by users, the system is able to propose and present, automatically and in real time, the matches between demand and supply of those intangibles (interests/ expertise/ know-how) that all companies would like to sell.

The result of the activity converges in all the networks that intends to integrate in the system (Facebook, Yammer, LinkedIn or Xing) and where the user is already connected to.

The user explicitly provides the system with useful information for refining the profile in order to put in evidence his interest in the received information.

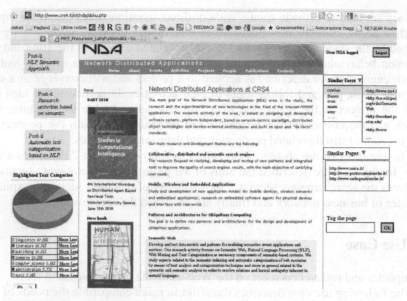

Fig. 1 The content visualization

The user can access his profile to verify if it is really consistent or if a distorted image of his interests is emerging. The system is able to state the reason why a specific correspondence has been proposed.

3 Goals of the Project

With social networks, blogs, RSS and new features in search engines are all news in the ICT context if compared with some years ago. This changed the way to access the information. The trend, hopefully, is the definition of new tools developed in order to follow the user in his activity and support him with the automatic generation and delivery of contents without his explicit request and according to his interests.

The automatic categorization of information through a predefined taxonomy, organized in a hierarchical category system, is often a restrictive and forced path. The same resource could be classified in different ways from different people and the same user could place the same page under different categories according to the reading context or to the content of interest. The classification of a document is, as well, depending on the personal culture, experience and context of life. Moreover, documents are often achieved using heterogeneous contents, talk about several topics and are obviously related to several categories.

Otherwise, with the Web 2.0, folksonomy, social tagging and social bookmarking place the user as starting point in a categorization work where each user labels resources. This step moves from a hierarchical logic to a more simpler way where all tags are at the same level.

Moving from the user management of information to an automatic one, a classification system should be able to categorize information according to user preferences and to relate his classification to a common set of categories based on a predefined taxonomy. By means of a such categorization tool, each user manages in a personal way his bookmarks, accedes to a quantity of Web sites, about scientific, news, entertainment or other topics, selecting, choosing and categorizing through the system. The system has to be able to manage a flow of data coming from a big set of predefined channels and updatable depending on the user preferences. Channels should be social networks, blogs, RSS services, news services, Web sites and also search engines, selected by the user.

To categorize information from these channels and to deliver contents that meet user preferences by means of a match algorithm based on the user profile and the document classification is a crucial point. The user can see categories associated to each resource labelled and ordered according to his schema.

4 A Semantic Search Engine for a Business Network

The semantic search engine is intended as a support tool for users, an active assistant able to give in *real time* references for the use of the information, reporting as more interesting the information that might match with the personal interest specified in the user profile.

There are two kinds of users: *companies* and the *generic user*, including employees, researchers, professionals, and people having specific interests and skills, according to the resources associated with them and emerging by their daily activities, that the system is able to track.

The business network is a point of contact between the academic and research world in general and the productive one and defines a communication level between users belonging to a community. The business network facilitates the sharing of knowledge, ability, expertise, skills, interests and resources between users belonging to the community that need or are interested in specific topics. In fact, it is not always easy to rise these feature, especially the immaterial expertise. But even publications or ongoing or past projects in which someone is involved, are often dispersed between public databases, or can be found only in the intranct of each company, or sometimes exists only in the head of someone, and it is not easy to explicit them. All the members of the community are linked together by the net of their skills: they are both depository of expertise in the service of users who need it: on the other hand they can need skills (papers, suggestions, projects, contacts) that other members can make available. This can be achieved with the development of an application, running on the computer of the user, that filters his activities and modifies his status, walls and links of the social network that the user subscribed, according to his permissions. Simultaneously records the activities on the user database.

The aim is to encourage the cooperation and the sharing of ideas, of different point of views, information material or needs, and to support the productive world and decision-making connected with it.

5 The Client Application

Users are organized as a community, configured according to their activities, through the management and by reporting organized content, information dynamically updated and personalized according to the specific user profile. The system will provide access to sources of shared documentation, to monitoring data, to support tools for sharing information between users, to networks of contacts explicitly specified in the community. The application shares this data with the other users that subscribed the community so that each user, according to the settings and the permissions, should know which resources have been visited, from whom and when. The application communicates these information to a plug-in installed on the user's browser that alerts the user and updates the visualization of information according to his preferences.

The system manages the user profile in order to control how the user preferences evolve during sessions of work. Information is monitored at time interval and new sessions can modify user preferences.

The system starts with a predefined user profile and evolves subsequently, using text categorization tools in order to categorize resources that are actually read, saved, commented. Only in these cases the system will modify the user criterion of classification for subsequently analysis.

The system follows step by step the evolution of user interests and suggests him, through the analysis of his profile, topics of interest, documents, contacts, etc, according to his interests. Moreover, the system is able to associate user profiles to companies or project profiles, automatically generating in real-time networks of expertise based on several configurable parameters and requirements.

Figure 2 shows a general description of the client application and the flow of the data coming from several distributed sources, such as social networks, blogs, RSS pages, visited Web pages, etc. There are four main modules: the User Profiling Module, the Collaborative Filtering and Recommendation System Module, the Classifier and the Matching Module, each responsible of the functionalities described below.

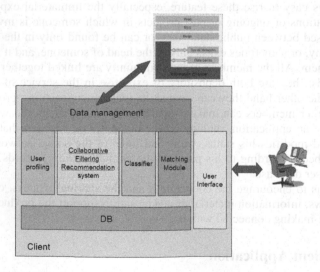

Fig. 2 The client application

The level of communication between the modules and the distributed information is regulated by a layer that receives the data coming from the sources, and after an analysis and an opportune elaboration, is able to deliver to each module the portion of information that they are able to manage. Each module performs his activity, sometimes collaborating with other modules, and the result of the process is saved on a database. The interface allows queries in natural language and presents results according to the user profile and preferences. The system will be able to retrieve information from several textual and multimedia sources, and from Web services, even if conditionally.

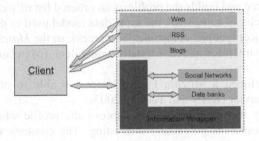

Fig. 3 The information wrapper

Figure 3 shows in a summary way the data sources and a module named Information Wrapper that uniforms data coming from data banks (DBLP, ACM DL sites or institutional databases) and, under particular conditions, from social networks.

Some sources such as news services, social networks, blogs, RSS feeds, will be selected by the user or they will be automatically proposed by the system, by means of the preferences expressed by default or defined by the user profile and by the interests identified by the viewed pages.

Other content will consist of personal and corporate profiles extracted from the HTML home pages, abstracts of scientific publications, bibliographies extracts from data banks.

More details of the modules involved in the system are described below.

5.1 Search Engine

The search engine module is contained in the Data Management Module, still under definition. The search engine indexes information coming from data sources and manages information related to the users, communities, companies, events, etc.

5.2 User Profiling

User profiling is a crucial process of the system because it has to define the user's interest, allowing the collaborative filtering and the recommendation tools to select and send information useful for the user itself (Burke R, 2002).

The *User Profile Module* (UPM) is able to manage user information. It involves the *Classifier Module* to analyze the textual data available on resources which the user interacts with, during his usual Web navigation activities.

Web sites, Social networks and Internet communities are an excellent source of information about users' interests. The system uses them as starting point of the creation of a user profile. The Web pages of personal and company site, the presented articles of the reported blogs and the curriculum vitae, are processed by the *Classifier Module* that returns the categories related to every resource above mentioned and provides to the UPM the first set of data to create the user profile. The module, according to user's preferences and number of resources, assigns a weight to every data source and builds the profile as an ordered list of weighted categories based on a hierarchy of categories, the same data model used to describe every resource in this project (people, companies, resources), so the *Matching Module* will be able to compare information formally equivalent (Marcialis and De Vita, 2008).

We have developed a Mozilla Firefox extension able to observe and store user's surfing behavior (Kelly and Teevan, 2003).

The UPM uses the same process to improve the profile when the extension points out new resources potentially interesting. The contents viewed, the time spent and the actions performed (printing, downloading, bookmarking, etc.) on each Web page the user visits are some of the information that the extension manages; it, handling the DOM of the page, also offers to the user tools to: mark a content, or a portion of it, as interesting; tag a textual content using a hierarchical set of categories; categorize a text portion and to modify this categorization; point out which results, among the ones offered by a search engine, can satisfactorily solve the query and ask questions related to a content to the user community.

The extension processes all these data, and notifies the UPM about the interesting resources identified.

A profile for both users, companies and researchers, is defined creating in such a way a history depending on their activities and behavior. So, the system will attempt to identify user requirements and to predict its future behavior and interests, in order to automatically propose resources useful to its activities without the need to search for them.

Data collected in this way are used by the system to find similarities, complementarities and links between companies and researchers, thus facilitating the match between supply and demand, particularly for intangibles such as interest, expertise, know-how.

The user should be able to access to its profile in order to check the reliability of the image that the system is bringing out, providing a positive or negative feedback to the matching proposed by the system and should manually change categories or preferences he doesn't agree.

5.3 Collaborative Filtering

During his activities, the user is supported by a module that helps him through two very important features: a collaborative filtering (De Vita et al., 2008) and a recommendation system. This module filters information by means of parameters based on the user's preferences and his profile and gives advice to the user for news regarding communities and network activities that should be of interest. Advices are about new activities performed by users having similar interests, companies having similar profile or by researchers having similar profile (based on their curriculum vitae). Web contents are also highlighted: events of the network like workshop and conferences; documents, papers, notes, projects, reviews classified that match users interests or announcements of competition, calls, etc.

By means of the indications given by the user to the system it is possible to refine the profile.

During the Web navigation the extension's functionalities allows the user to see highlighted contents when they are identified as interesting from the user community. The extension also highlights all the search engine results recommended by users having submitted a similar query and the ones that link resources with an interesting content. Moreover the extension augments the displayed resource with additional information, by means of Post-it notes, the users community leaves on the page and links to similar contents recommended by the same community. The extension also allows users to ask a question related to a portion of page or, if he wants, to answer questions of users community.

5.4 Data Categorization

The system, with the user profile module, compares user profiles to company profiles through data categorization. It matches similar profiles, compares curricula of the user with request coming from companies, filters news and contents coming from the search engine working on the semantic of texts. More in details, the semantic analysis of texts let the system to create a structure able to give the right meaning to groups of words according to their occurrence within text and solving misunderstandings related to thesaurus, slang expressions and meaning, improving the information retrieval and the knowledge management. Semantic text categorization techniques allow to classify texts into one or more predefined categories, according to the meaning expressed in their content. Although several classifications techniques and accurate methods to classify text documents are available, such as statistical approaches or systems that automatically classify text documents into predefined thematic classes, we prefer a classifier able to categorize a great variety of resources, different in content, type, format, in language and vocabulary, that are not always mapped on taxonomies or ontologies. In fact it is not always easy to map generic resources to taxonomies or to find a corpus of tagged resources to categorize them by means of traditional categorization techniques. To address all these problems, we adopted a semantic approach to develop a categorizer, i.e., a module able to manage resources and queries by means of semantic text categorization techniques.

The classifier is based on a hierarchy of categories proposed by WordNet Domains (Magnini et al., 2002). These categories are the set of starting used by the system for the text categorization of resources.

The classifier performs a semantic disambiguation through the identification of relation between terms in order to identify composed terms, word sense disambiguation, name entities, geographic location.

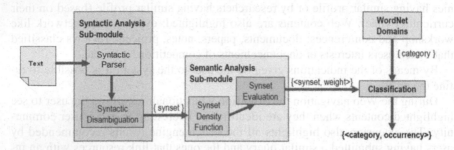

Fig. 4 The schema of the Classifier

As illustrated in Figure 4, the main phases after the parsing of the text of resources (Web pages, documents, notes, etc) follows the analysis and the syntactic disambiguation (Sleator and Temperley, 1993) (Liu, 2004) of the sentences. This step is performed by means of syntactic parser in order to identify the part of speech in phrases and give to the semantic sub-module the more relevant nouns, verbs, adjectives and adverb extracted from each phrase.

By means of a density function (Addis et al., 2009), the semantic disambiguation of a sentence allows to assign to each term its most likely meaning in the context of the phrase, choosing between all their senses available from WordNet. The WordNet dictionary groups nouns, verbs, adjectives, adverbs and organizes them into synonyms sets, called synsets, each expressing a distinct concept uniquely identified by a synsetID. In WordNet synsets are linked by means of conceptual-semantic and lexical relations, such as synonymy, meronymy/holonymy, hyperonymy/hyponymy, etc. The goal of the semantic disambiguation task is to reduce the number of synsets that have been activated by the syntactic analysis, evaluating the use of words in the context of a sentence and the use of sentences in the context of the document. We calculate the synset density evaluating how many times a synset is included in a document, with respect to all the synsets of the document. In so doing, we can take advantage of the synonymy relation available from WordNet to identify within phrases similar terms with the same sense. Moreover, by means of word sense density, we can categorize the whole document, assigning to it some domain topics during the semantic disambiguation phase, identifying all the synsets referred to the content of a document, and evaluating its most probable sense.

The semantic sub-module identifies name entities and geographic locations and geo-referencing of resources.

Finally, a classifier sub-module performs the categorization of the textual resource by categories and values (Angioni et al., 2008a). The classifier is capable to categorize documents automatically, applying a classification algorithm based on the Dewey Decimal Classification, as proposed in WordNet Domains. The results set of categories is further reduced by the application of a function that takes into account only categories characterized by a density value bigger than a fixed range value.

5.5 Matching Module

The module is responsible to perform the matching between the information coming from the several data sources and by the users' profile, identifying those of real interest for each user.

It is able to organize data coming from users and companies' profile, managing the textual resources, such as notes, papers, comments, profile data, previously analyzed by the classifier and aggregate the information.

Finally it sends notifications to users and the information as elaborated by the specific algorithm of matching.

We use the same model to describe users and resources, an ordered list of weighted categories. Using a similarity algorithms is possible to compare numerically users, pages, researchers and companies. Moreover, calculating distance between users' profiles, the Qualified Users Set (QUS) is identified. The QUS is a restricted set of users similar to the active user, able to recommend resources to the active one.

6 Conclusion

The Web is changing and the way to access the information and the contents themselves are evolving too. Social networks, blogs, rss and new users' supports based on NLP are defining new evolutionary scenarios and creating new expectations for the Web. In this chapter we illustrated a project named *A Semantic Search Engine for a Business Network* that defines a scenario where the above tools will converge in a system that, in our intention, will implement the use case described as a step of the vision described in the chapter. The approach described aims at the development of the business network between the academic and the research world and the productive one, allowing a point of contact between users putting in evidence theirs skills and expertises.

The project aims both at implement the features described and at define and implement the described scenario. A validation to support the value of the expressed ideas will be one of the goal of the above mentioned project, where experimental results will be product.

References

Addis, A., Angioni, M., Armano, G., Demontis, R., Tuveri, F., Vargiu, E.: A Novel Semantic Approach to Create Document Collections. In: dos Reis, A.P. (ed.) Proceedings of Intelligent Systems and Agents, pp. 53–60. IADIS Press (2008); Selected for the best paper award

Angioni, M., Demontis, R., Tuveri, F.: A Semantic Approach for Resource Cataloguing and Query Resolution. Communications of SIWN. Special Issue on Distributed Agent-based Retrieval Tools 5, 62–66 (2008a)

Burke, R.: Hybrid Recommender Systems: Survey and Experiments. In: User Modeling and User-Adapted Interaction, vol. 12(4), pp. 331–370. Kluwer Academic Publishers, Hingham (2002)

De Vita, E., Deriu, M., Marcialis, I., Paddeu, G.: Personalization and Collaborative Filtering for Information Retrieval on the Web. Communications of SIWN. Special Issue on Distributed Agent-based Retrieval Tools 5(-), 51–56 (2008)

Marcialis, I., De Vita, E.: SEARCHY: An Agent to Personalize Search Results. In: Mellouk, A. (ed.) Third International Conference On Internet And Web Applications And Services, pp. 512–517 (2008); IARIA. Institute of Electrical and Electronics Engineers (IEEE). Authorized distributor of all IEEE proceedings

Horowits, D., Kamvar, S.: The Anatomy of a Large-Scale Social Search Engine. Submitted to WWW 2010, Raleigh, NC, USA (2010)

Kelly, D., Teevan, J.: Implicit Feedback for Inferring User Preference: A bibliography. SIGIR Forum. 37(2), 18–28 (2003)

Liu, H.: MontyLingua: An end-to-end natural language processor with common sense (2004), http://web.media.mit.edu/~hugo/montylingua (viewed March 30, 2010)

Magnini, B., Strapparava, C., Pezzulo, G., Gliozzo, A.: The Role of Domain Information in Word Sense Disambiguation. Natural Language Engineering, special issue on Word Sense Disambiguation 8(4), 359–373 (2002)

Mislove, A., Gummadi, K., Druschel, P.: Exploiting Social network for Internet Search. In: Proceedings of the 5th Workshop on Hot Topics in Networks, Irvine, CA (2006)

Sleator, D.D., Temperley, D.: Parsing English with a Link Grammar. In: Third International Workshop on Parsing Technologies (1993)

Spivack, N.: Eliminating the Need for Search-Help Engines (2010), http://www.novaspivack.com/uncategorized/eliminating-the-need-to-search (viewed March 30, 2010)

Wissner, J., Spivack, N.: Case Study: Twine. In W3C, Semantic Web Use Cases and Case Studies (2009), http://www.w3.org/2001/sw/sweo/public/UseCases/Twine (viewed March 30, 2010)

A Collaborative Web Application for Supporting Researchers in the Task of Generating Protein Datasets

Giuliano Armano and Andrea Manconi

Abstract. The huge difference between known sequences and known tertiary structures has fostered the development of automated methods and systems for protein analysis. When these systems are learned using machine learning techniques, the capability of training them with suitable data becomes of paramount importance. From this perspective, the search for (and the generation of) specialized datasets that meet specific requirements are prominent activities for researchers. To help researchers in these activities we developed ProDaMa-C, a web application aimed at generating specialized protein structure datasets and fostering the collaboration among researchers. ProDaMa-C provides a collaborative environment where researchers with similar interests can meet and collaborate to generate new datasets. Datasets are generated selecting proteins through user-defined pipelines of methods/operators. Each pipeline can also be used as starting point for building further pipelines able to enforce additional selection criteria. Freely available as web application at the URL http://iasc.diee.unica.it/prodamac, ProDaMa-C has shown to be a useful tool for researchers involved in the task of generating specialized protein structure datasets.

1 Introduction

The steady work of researchers in conjunction with the advances in technology (e.g. high-throughput technologies) has arisen in a growing amount of known sequences. This rapid growth has been accompanied by a huge increase of biological databases (Fig. 1), which has not been accompanied by an equally rapid growth of the known

Giuliano Armano
Dept. of Electrical and Electronic Engineering, University of Cagliari, Italy
e-mail: armano@diee.unica.it

Andrea Manconi
Institute for Biomedical Technologies, National Research Council, Milano, Italy
e-mail: andrea.manconi@itb.cnr.it

V. Pallotta, A. Soro, and E. Vargiu (Eds.): Advances in DART, SCI 361, pp. 13–26.
springerlink.com   © Springer-Verlag Berlin Heidelberg 2011

structures. In fact, as the genome projects worldwide progress, the difference between known sequences and known tertiary structures increases exponentially. This
is particularly evident when comparing the yearly deposited sequences in GenBank
[1] (Fig. 2) with the yearly released structures in the Protein Data Bank (PDB) [2]
(Fig. 3). The actual rate between the former and the latter is currently over three
order of magnitude.

This huge discrepancy has fostered the development of automated methods and
systems for protein analysis. To cite a few examples, let us recall systems to predict
protein secondary structure (e.g., PSIPRED [3] and SSPRO [4][5]), to predict transmembrane regions (e.g., TMBpro [7]), and to predict beta-turns (e.g., BTPRED [8]
and [9]).

Most of the systems designed to perform protein analysis make use of machine
learning techniques, so that the capability of training them with suitable data becomes of paramount importance. From this perspective, the search for and/or the
generation of specialized datasets that meet specific requirements are prominent activities for researchers. Different protein datasets have been proposed in the literature. However, these datasets are designed to investigate specific problems and may
not be in accordance with the needs of researchers, or may not fit the specific nature
of the problem in hand. Owing to these limitations, researchers are often involved in
the task of generating protein datasets. This task involves the problems of *i*) searching for, retrieving, and combining protein data from relevant specialized databases,
ii) preprocessing and analyzing these data with suitable tools, and *iii*) overcoming

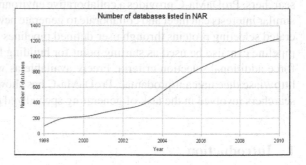

Fig. 1 The number of resources listed from 1998 to
2010 in the Nucleic Acids
Research database.

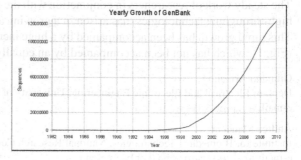

Fig. 2 The number of sequences deposited in Gen-
Bank from December 1982
to August 2010 (GenBank
release 179.0).

the limitations associated with the migration of data and with the methods available for managing them.

- *Searching for, retrieving, and combining protein data.* Biological sources are considered strongly heterogeneous for many aspects, including:
    - *Representational heterogeneity.* Biological sources may represent data of the same kind in different ways. These data are typically represented in a variety of formats, and may be organized in flat files, relational or object-oriented databases. This kind of heterogeneity consists of structural, naming, semantic, and content differences [10].
    - *Autonomous and web based sources.* Due to their autonomy, biological sources are free to modify their design or schema, or to remove data without any warning to end users.
    - *Different querying capabilities.* Each source defines and provides different query interfaces, and can optionally restrict the access to its data. These restrictions force end-users and external systems to adapt their queries.
- *Preprocessing and analyzing biological data.* These activities are very important for researchers engaged in the task of generating specialized datasets. Protein data retrieved from different sources need to be preprocessed and analyzed to be used. In this context, tools for protein identification and characterization, similarity and identity searches, pattern and profile searches, and structure analysis are commonly used by researchers.
- *Migrating data.* Due to the complexity of preprocessing and analysis, the use of multiple specific tools is typically required. Researchers often concentrate on how to overcome the problems related with the migration of data from several different tools.

Several approaches and systems have been proposed in the literature to deal with the problem of integrating biological sources. WWW-Query [11] is a SRS (sequence retrieval system) [12] that integrates the ability of retrieving relevant protein sequences with tools for analyzing them. The associated tools consist of multivariate methods for studying codon and amino acid composition and for complementing phylogenetic analysis. TAMBIS [13] (Transparent Access to Multiple Bioinformatics Information Sources) is a mediator-based system for performing bioinformatics

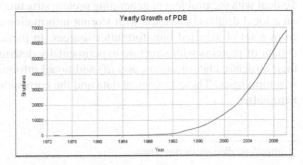

Fig. 3 The number of searchable structures per year in the Protein Data Bank.

tasks over multiple information systems with an interactive user interface. It is based on a model of the concepts and their relationships in molecular biology and bioinformatics. k2/BioKleisli [14] is a mediator-based system to access data sources critical to the human genome project. It is intended to provide (read-only) access to multiple data sources characterized by complex and structured data. Biopython [15] is a Python library aimed at helping reserchers in the task of managing bioinformatics data. Mainly aimed at parsing bioinformatics files into Python data structures, it is able to cope with a set of popular on-line bioinformatics sources, and to provide interfaces to commonly-used programs. Although several approaches and systems have been proposed, an approach able to solve all mentioned problems does not exist yet.

To generate protein structure datasets, major benefits can be obtained with a set of specialized tools for automatically retrieving and organizing relevant protein data, as well as analyzing and selecting them according to specific, application-dependent, constraints. In our opinion, the social and collaborative nature of the Web 2.0, which encourages data integration as well as data sharing and reuse, is expected to provide a significant contribution to overcome the problems previously mentioned. In this chapter, we propose and describe ProDaMa-C (*Protein Dataset Management - Collaborative*), a collaborative web application aimed at generating and sharing specialized protein structure datasets. Freely available for non-commercial use at http://iasc.diee.unica.it/prodamac/, ProDaMa-C has been developed using ProDaMa[1] [16], a library of Python APIs devised to provide full support for generating protein structure datasets. The remainder of the chapter is organized as follows. First a detailed description of the biological data integrated in ProDaMa-C is provided. A description of the methods available off-the-shelf for generating datasets follows. Then the benefits brought by ProDaMa-C are discussed. Conclusions end the chapter.

2 Biological Sources

Advances in high-throughput technologies for DNA sequencing and genomics has arisen in a rapid and huge increase of biological data. Several types of data are provided by biological sources. Genome, protein-structure, protein-model, protein-protein, and microarray databases are types of data exported by biological sources. To deal with the problem of generating protein structure datasets, ProDaMa-C relies on a local database entrusted with storing information about protein data retrieved from a set of selected bioinformatics sources. In particular ProDaMa-C contains: *i*) data of proteins with structure experimentally determined, *ii*) information about their structure classifications, and *iii*) additional information about membrane protein topologies. These biological data and the corresponding sources are described hereinafter.

---

[1] http://iasc.diee.unica.it/prodama

2.1 Protein Structures

In ProDaMa-C, information about proteins with known structure is retrieved from the PDB. It is the single repository of information about the 3D structures of large biological molecules, primarily proteins and nucleic acids. It was founded in 1971 at Brookhaven National Laboratory. Improvement in technology have given rise to an exponential increase of the number of structures determined. About 69000 structures are currently stored in the PDB. The data contained in the archive are mostly obtained by X-ray crystal structure determination, NMR, and electron mycroscopy (Fig. 4).

The role of the PDB in representing the structures experimentally determined has become more and more important, insomuch as by the early 1990s the majority of journals required a PDB identification for publication.

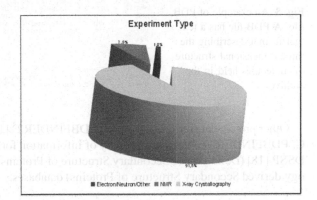

Fig. 4 The distribution of types of experiment used to determine the structures stored in the PDB (from the 2009 annual report).

These data are represented using a specific format, called the PDB file format (Fig. 5), which provides a standard representation for macromolecular structure data. This format consists of a collection of records that describe the atomic coordinates, chemical and biochemical features, experimental details related with the method used for determining the structure, and some structural features (e.g. the secondary structure assignments).

Historically, data were directly accessed as PDB files. Advances in the Internet technology promoted the development and the adoption of several ways for querying, accessing, and analyzing the data stored in the archive. Nowadays, the PDB website[2] integrates several tools (e.g. for structure visualization) and third-party databases (e.g. for structure classification) for analyzing and querying protein data. In addition, it provides several and specialized web-services aimed at querying and retrieving its data. However, the standard PDB file format remains the only way to access in a row all information related with a PDB entry.

---

[2] http://www.rcsb.org

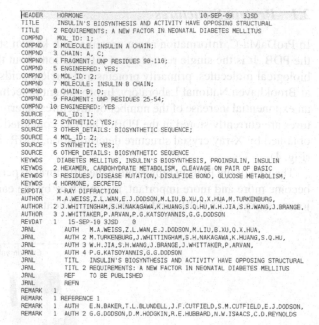

```
HEADER    HORMONE                                        10-SEP-09   3JSD
TITLE     INSULIN'S BIOSYNTHESIS AND ACTIVITY HAVE OPPOSING STRUCTURAL
TITLE    2 REQUIREMENTS: A NEW FACTOR IN NEONATAL DIABETES MELLITUS
COMPND    MOL_ID: 1;
COMPND   2 MOLECULE: INSULIN A CHAIN;
COMPND   3 CHAIN: A, C;
COMPND   4 FRAGMENT: UNP RESIDUES 90-110;
COMPND   5 ENGINEERED: YES;
COMPND   6 MOL_ID: 2;
COMPND   7 MOLECULE: INSULIN B CHAIN;
COMPND   8 CHAIN: B, D;
COMPND   9 FRAGMENT: UNP RESIDUES 25-54;
COMPND  10 ENGINEERED: YES
SOURCE    MOL_ID: 1;
SOURCE   2 SYNTHETIC: YES;
SOURCE   3 OTHER_DETAILS: BIOSYNTHETIC SEQUENCE;
SOURCE   4 MOL_ID: 2;
SOURCE   5 SYNTHETIC: YES;
SOURCE   6 OTHER_DETAILS: BIOSYNTHETIC SEQUENCE
KEYWDS    DIABETES MELLITUS, INSULIN'S BIOSYNTHESIS, PROINSULIN, INSULIN
KEYWDS   2 HEXAMER, CARBOHYDRATE METABOLISM, CLEAVAGE ON PAIR OF BASIC
KEYWDS   3 RESIDUES, DISEASE MUTATION, DISULFIDE BOND, GLUCOSE METABOLISM,
KEYWDS   4 HORMONE, SECRETED
EXPDTA    X-RAY DIFFRACTION
AUTHOR    M.A.WEISS,Z.L.WAN,E.J.DODSON,M.LIU,B.XU,Q.X.HUA,M.TURKENBURG,
AUTHOR   2 J.WHITTINGHAM,S.H.NAKAGAWA,K.HUANG,S.Q.HU,W.H.JIA,S.H.WANG,J.BRANGE,
AUTHOR   3 J.WHITTAKER,P.ARVAN,P.G.KATSOYANNIS,G.G.DODSON
REVDAT   1   15-SEP-10 3JSD    0
JRNL        AUTH   M.A.WEISS,Z.L.WAN,E.J.DODSON,M.LIU,B.XU,Q.X.HUA,
JRNL        AUTH 2 M.TURKENBURG,J.WHITTINGHAM,S.H.NAKAGAWA,K.HUANG,S.Q.HU,
JRNL        AUTH 3 W.H.JIA,S.H.WANG,J.BRANGE,J.WHITTAKER,P.ARVAN,
JRNL        AUTH 4 P.G.KATSOYANNIS,G.G.DODSON
JRNL        TITL   INSULIN'S BIOSYNTHESIS AND ACTIVITY HAVE OPPOSING STRUCTURAL
JRNL        TITL 2 REQUIREMENTS: A NEW FACTOR IN NEONATAL DIABETES MELLITUS
JRNL        REF    TO BE PUBLISHED
JRNL        REFN
REMARK   1
REMARK   1 REFERENCE 1
REMARK   1  AUTH   E.N.BAKER,T.L.BLUNDELL,J.F.CUTFIELD,S.M.CUTFIELD,E.J.DODSON,
REMARK   1  AUTH 2 G.G.DODSON,D.M.HODGKIN,R.E.HUBBARD,N.W.ISAACS,C.D.REYNOLDS
```

Fig. 5 An example of PDB file. A PDB file has a textual format describing the three dimensional structure of molecules held in the archive.

Other protein data retrieved from the PDBFINDER2[3] [17] are stored in ProDaMa-C. PDBFINDER2 contains a summary of information for PDB entries extracted by DSSP [18] (Definition of Secondary Structure of Proteins) and HSSP [19] (Homology derived Secondary Structure of Proteins) databases.

2.2 Structural Classifications

The structure of a protein can help researchers to make conjectures aimed at inferring its biological function and at understanding its evolution. A prerequisite for both is to know its structure and its relationships with other proteins. In this perspective, the availability of algorithms for structural alignment and structural classification has given a significant contribution to the understanding of evolutionary mechanisms. Several approaches for structure classification (e.g., SCOP [20], CATH [21], PFAM [22], DALI [23] [24]) have been proposed in the literature. ProDaMa-C embeds the information retrieved from the largest structural classification repositories, i.e., SCOP and CATH. SCOP[4] (Structural Classification of Proteins) uses for classification a method based on visual inspection of structures first compared using automatic procedures. The classification is made according to hierarchical levels (i.e., class, fold, superfamily, and family), which embody evolutionary and structural relationships. CATH[5] is a (semi-automatic) hierarchical domain classification of protein

---

[3] http://swift.cmbi.ru.nl/gv/pdbfinder/

[4] http://scop.mrc-lmb.cam.ac.uk/scop/index.html

[5] http://www.cathdb.info/

structures. Only crystal structures with a resolution better than 4.0Å are considered, together with NMR structures. The classification procedure uses a combination of automated and manual techniques which include computational algorithms, empirical and statistical evidence, literature review and expert analysis. There are four levels of classification in the defined hierarchy: (C)lass, (A)rchitecture, (T)opology, and (H)omologous superfamily. The structural classifications of SCOP and CATH can be obtained as flat files from the corresponding websites. However, wrapping their websites is the most suitable way for querying and retrieving data from both. In this way one is sure to access the latest release of the databases.

2.3 Membrane Proteins

A membrane protein is a protein molecule attached to, or associated with, the membrane of a cell. These proteins cover a very important role in the communication between cells and perform a wide range of biological functions, such as respiration, signal transduction and molecular transport. Several membrane protein databases have been proposed in the literature. The reliability of the data concerning this kind of proteins is very important for studying them. In a survey performed on SWISS-PROT [26], Senes et al. [25] determined that almost 94% of the transmembrane segments, instead of being experimentally derived, were in fact annotated as potential, possible, or probable, as they were identified by means of prediction techniques, primarily hydropathy plots. In this perspective, ProDaMa-C integrates the data about membrane proteins retrieved from MPtopo[6] [27], a database of membrane proteins whose topologies have been verified experimentally by means of crystallography, gene fusion, and other methods. MPtopo data can be obtained as a flat file or by querying the website. Also in this case the most suitable way for querying and retrieving relevant data is wrapping the website.

3 Methods

As already pointed out, ProDaMa-C relies on a local database entrusted with storing information retrieved by a set of specialized biological sources. The local database contains: *i*) protein data, retrieved from the Protein Data Bank, *ii*) information about their structural classification, retrieved from the SCOP and CATH databases, *iii*) information about membrane protein topologies, retrieved from the MPtopo database, and iv) other information (e.g. quality parameters), retrieved from the PDBFINDER2 database.

All locally-stored data are periodically updated according to the information contained in the corresponding repositories. A user agent looks for obsolete and new structures in the PDB, and the local database is updated accordingly.

The local database is also pre-loaded with a number of commonly used biological datasets, i.e., RS126 [28], PDBSELECT25 [29], the PDB clusters of structures

---

[6] http://blanco.biomol.uci.edu/mptopo/

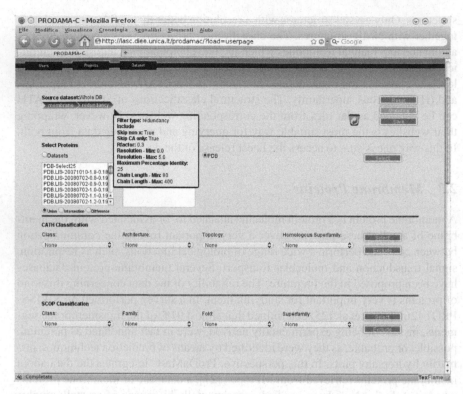

Fig. 6 A screenshot of the graphical user interface available for building pipelines.

based on 50%, 70%, 90% and 95% sequence identity, and the datasets of sequence structures used by WHAT IF [30] based on sequence identity, resolution and R-factor.

In ProDaMa-C, a new dataset is generated by applying a user-defined pipeline of methods and operators to the local database or to a previously-generated dataset. Furthermore, an existing pipeline can be used as a starting point to build new pipelines by adding further constraints or modifying existing ones. To generate a pipeline, three groups of operators are available off-the-shelf.

- *Search methods* are typically applied to select proteins that satisfy homology and/or similarity constraints. In particular, FASTA [31] and PSI-BLAST services are available, useful to perform search by sequence similarity. Search by sequence identity is performed by querying PISCES [32]. Methods for SCOP and CATH protein similarity search and for transmembrane protein topology search are also provided. Furthermore, proteins can be selected by imposing constraints on their quality –i.e., on the experimental method that has been used, on the X-ray resolution, as well as on their R-factor and free R-factor.

- *Filter operators* are aimed at selecting relevant proteins according to a unary or binary predicate. In particular, given the input dataset, it is possible to select

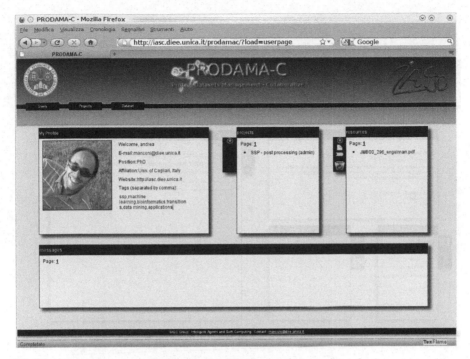

Fig. 7 A screenshot of the personal dashboard associated to a registered user.

proteins according to the following constraints: i) single vs. multiple chains; ii) sequence length, e.g., length ≤ 200; iii) protein structure, such as the number and/or length of alpha-helices or beta-strands or the number of transmembrane segments; iv) percent of identity, e.g. sequence identity $\leq 25\%$.

- *Set operators* provide the classical union, intersection, and difference.

According to the Web 2.0 philosophy, ProDaMa-C has been devised to promote the collaboration among researchers with similar interests. To this end, it supports the creation of workgroups involved in common projects. Registered users can access all services provided by ProDaMa-C. In particular, users can i) build and modify their profile, ii) manage a personal knowledge repository, iii) create new projects, and iv) join projects created by other users. Each user has an associated personal dashboard (Fig. 7) to manage pipelines and related databases, projects (including membership in other projects), and documents uploaded to the system.

As already pointed out, users with similar interests can also create groups to work on common projects. Each project has an associated knowledge repository (Fig. 8) where researchers can share ideas, documents, datasets, as well as pipelines useful to generate datasets in accordance with their needs. To take advantage of the collaborative facilities provided by ProDaMa-C, a user can characterize her/himself with suitable tags (free-text keywords). The same tag oriented strategy can also be used to categorize projects. In so doing, users can easily discover other users and/or

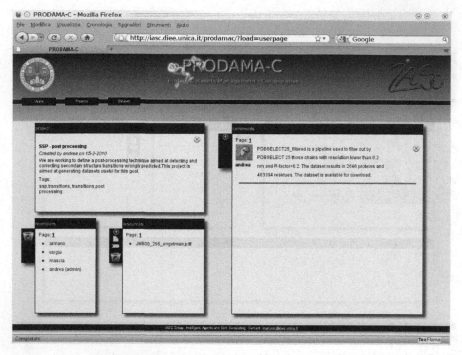

Fig. 8 A screenshot showing the web area associated to the knowledge repository of a project.

relevant projects related to their research interests. Anonymous users have access to a limited set of functionalities. In particular, although with full access to all project resources, they cannot update them.

4 Discussion

Devised and developed to facilitate the work of researchers engaged to generate specialized protein strucuture datasets, ProDaMa-C is a tool able to address a wide range of problems. We deem that ProDaMa-C can be very effective in the task of generating datasets according to specific constraints on protein structure. In particular, ProDaMa-C can be used to generate datasets according to constraints on the protein quality parameters, structure composition, structure classifications, and membrane topologies.

Several experiments have been performed, aimed at assessing the ability of ProDaMa-C to generate and handle specific datasets, while speeding up the whole process and also thanks to a user-friendly interface. To cite few examples of dataset generation, ProDaMa-C made it easier to generate: *i*) the dataset used for training and testing the YASPIN [33] prediction method, where a non-redundant set of proteins with known structures has been obtained removing all transmembrane entries (SCOP class f) from PDBSELECT25; *ii*) the dataset for training the JNET [34] secondary structure prediction method, where the starting set CB513 [35] of proteins with known strucutures has been screened to remove those shorter than 30 residues, and those from families that contained only two sequences; *iii*) the RS126 dataset, obtained by imposing constranints on the pairwise sequence identity ($\leq 25\%$) and on the sequence length (≥ 80 residues); *iv*) the dataset for training the β-turn prediction method BTPRED, obtained selecting the structurally non-homologous proteins according to the CATH classification, with less than 25% pairwise sequence identity, and resolution 2Å or better; and *v*) the dataset used in [36] for analysing two state-of-the-art secondary structure prediction methods, PHD [37][38] and JPRED [39], where the structures in the dataset PDBSELECT25 are restricted to those solved by X-ray crystallography, with a maximum R-factor of 0.2, and a minimum helical content of 10%.

These experiments highlighted that, thanks to to the integrated filters and methods, ProDaMa-C is well-versed to tackle with several problems. The ability to speed up the process of generating datasets is mainly due to the local database and to the proposed approach based on pipelines. While the availability of protein data, periodically and automatically updated, lightens the users from the tasks of searching for, retrieving, and preprocessing suitable data, the pipeline-based approach allows them to quickly apply changes to existing pipelines (e.g. to modify a pipeline according to new criteria) and to use existing pipelines with a different starting set of proteins (e.g. to update a dataset according to new protein data stored in the system).

To help researchers in the task of building pipelines, a specialized user interface is available (Fig. 6), equipped with a simple graphic editor.

5   Conclusions

Protein sequence analysis is an important research area in bioinformatics, owing to the huge difference between the number of known sequences and known tertiary structures. This discrepancy has promoted the development of automated methods of analysis. The accuracy of these systems being related to the data used in the training phase, researchers are often involved in the task of searching for and/or generating specialized protein datasets. This task is very difficult and time-consuming as it typically requires to search for, retrieve and combine protein data, preprocess and analyze them with suitable specialized tools, while overcoming the limitations concerning the migration of data. Despite the huge increase of biological data and sources, the distinctive feature in integrating biological data is mainly concerned with the complexity of available data rather than with their amount [41].

In the literature several and specialized solutions have been proposed to deal with the problem of integrating biological sources. However, a standard approach able to solve the problem of integrating biological data has not been defined so far, and valid solutions aimed at helping researchers engaged in the task of generating protein structure datasets are not yet available. We deem that major benefits can be obtained by a specialized Web 2.0 application that integrates biological tools and data. In this perspective, we devised and developed ProDaMa-C, a collaborative web application designed to generate protein stucture datasets according to user-defined criteria, and to support the collaboration among researchers. To assess the potential of ProDaMa-C, we used it to generate several well-known datasets, taking care of enforcing the same constraints reported in the literature. Experimental results show that ProDaMa-C is able to generate specific datasets for a wide range of bioinformatics problems, while speeding up the whole process, also thanks to a user-friendly interface.

Acknowledgements

This work has been supported by the MIUR FIRB ITALBIONET (RBPR05ZK2Z), Bioinformatics analysis applied to Populations Genetics (RBIN064YAT_003) and SHIWA Projects. A special thanks goes to Stefano Curatti, for his help in implementing the web application.

References

1. Benson, D.A., Boguski, M.S., Lipman, D.J., Ostell, J., Ouellette, B.F.F., Rapp, B.A., Wheeler, D.L.: GenBank. Nucleic Acids Research 27(1), 12–17 (1998)
2. Berman, H.M., Westbrook, J., Feng, Z., Gilliland, G., Bhat, T.N., Weissig, H., Shindyalov, I.N., Bourne, P.E.: The Protein Data Bank. Nucleic Acids Research 28, 235–242 (2000)
3. Jones, D.T.: Protein secondary structure prediction based on position-specific scoring matrices. Journal of Molecular Biology 292, 192–202 (1999)
4. Pollastri, G., Przybylski, D., Rost, B., Baldi, P.: Improving the prediction of protein secondary structure in three and eight classes using recurrent neural networks and profiles. Proteins 47, 228–235 (2002)
5. Cheng, J., Randall, A., Sweredoski, M., Baldi, P.: SCRATCH: a Protein Structure and Structural Feature Prediction Server. Nucleic Acids Research Web Server Issue 33, 72–76 (2005)
6. Altschul, S., Madden, T., Schaffer, A., Zhang, J., Zhang, Z., Miller, W., Lipman, D.: Gapped BLAST and PSI-BLAST: a new generation of protein database search programs. Nucleic Acids Research 25(17), 2289–3402 (1997)
7. Randall, A., Cheng, J., Sweredosk, M., Baldi, P.: TMBpro: secondary structure, β-contact and tertiary structure prediction of transmembrane β-barrel proteins. Bioinformatics 24(4), 513–520 (2008)
8. Shepherd, A.J., Gorse, D., Thornton, J.M.: Prediction of the location and type of β-turns in proteins using neural networks. Protein Science 8, 1045–1055 (1999)
9. Kaur, H., Raghava, G.P.S.: Prediction of beta-turns in proteins from multiple alignment using neural network. Protein Science 12, 627–634 (2003)
10. Sujansky, W.: Heterogeneous database integration in biomedicine. Journal of Biomededical Informatics 34(4), 285–298 (2001)
11. Perrire, G., Gouy, M.: WWW-query: An on-line retrieval system for biological sequence banks. Biochimie 78(5), 364–369 (1996)
12. Etzold, T., Argos, P.: SRS – an indexing and retrieval tool for flat file data libraries. Bioinformatics 9(1), 49–57 (1992)

13. Stevens, R., Baker, P., Bechhofer, S., Ng, G., Jacoby, A., Paton, N.W., Goble, C.A., Brass, A.: TAMBIS: Transparent Access to Multiple Bioinformatics Information Sources. Bioinformatics 16(2), 184–186 (2000)
14. Davidson, S.B., Overton, C., Tannen, V., Wong, L.: BioKleisli: a digital library for biomedical researchers. International Journal on Digital Libraries 1(1), 36–53 (1997)
15. Chapman, B., Chang, J.: Biopython: Python tools for computational biology. ACM SIG-BIO Newslett. 20, 15–19 (2000)
16. Armano, G., Manconi, A.: ProDaMa: an open source Python library to generate protein structure datasets. BMC Research Notes 2, 202 (2009)
17. Hooft, R.W.W., Sander, C., Scharf, M., Vriend, G.: The PDBFINDER database: a summary of PDB, DSSP and HSSP information with added value. Bioinformatics 12(6), 525–529 (1996)
18. Kabsch, W., Sander, C.: Dictionary of protein secondary structure: pattern recognition of hydrogen-bonded and geometrical features. Biopolymers 22(12), 2577–2637 (1983)
19. Schneider, R., de Daruvar, A., Sander, C.: The HSSP database of protein structure-sequence alignments. Nucleic Acids Research 25(1), 226–230 (1997)
20. Andreeva, A., Howorth, D., Chandonia, J.M., Brenner, S.E., Hubbard, T.J.P., Chothia, C., Murzin, A.G.: Data Growth and its Impact on the SCOP Database: new Developments. Nucleic Acids Research 36, D419–D425 (2008)
21. Cuff, A.L., Sillitoe, I., Lewis, T., Redfern, O.C., Garratt, R., Thornton, J., Orengo, C.A.: The CATH classification revisited – architectures reviewed and new ways to characterize structural divergence in superfamilies. Nucleic Acids Research 37, D310–D314 (2009)
22. Bateman, A., Birney, E., Durbin, R., Eddy, S.R., Howe, K.L., Sonnhammer, E.L.L.: The Pfam Protein Families Database. Nucleic Acids Research 28(1), 263–266 (2000)
23. Holm, L., Sander, C.: Protein structure comparison by alignment of distance matrices. Journal of Molecular Biology 233(1), 123–138 (1993)
24. Holm, L., Rosenstrm, P.: Dali server: conservation mapping in 3D. Nucleic Acids Research 38, W545–W549 (2010)
25. Senes, A., Gerstein, M., Engelman, D.M.: Statistical analysis of amino acid patterns in transmembrane helices: the GxxxG motif occurs frequently and in association with β-branched residues at neighboring positions. Journal of Molecular Biology 296(3), 921–936 (2000)
26. Bairoch, A., Boeckmann, B., Ferro, S., Gasteiger, E.: Swiss-Prot: Juggling between evolution and stability. Briefings in Bioinformatics 5, 39–55 (2004)
27. Jayasinghe, S., Hristova, K., White, S.H.: MPtopo: A database of membrane protein topology. Protein Science 10, 455–458 (2001)
28. Rost, B., Sander, C.: Prediction of protein secondary structure at better than 70% accuracy. Journal of Molecular Biology 232, 584–599 (1993)
29. Hobohm, U., Sander, C.: Enlarged representative set of protein structures. Protein Science 3(3), 522–524 (1994)
30. Vriend, G.: WHAT IF: A molecular modeling and drug design program. Journal of Molecular Graphics 8, 52–56 (1990)
31. Pearson, W.R., Lipman, D.J.: Improved tools for biological sequence comparison. Proceeding of the National Academy of Sciences of the United States of America 85(8), 2444–2448 (1998)
32. Wang, G., Dunbrack, R.L.: Jr. PISCES: a protein sequence culling server. Bioinformatics 19, 1589–1591 (2003)
33. Lin, K., Simossis, V.A., Taylor, W.R., Heringa, J.: A simple and fast secondary structure prediction method using hidden neural networks. Bioinformatics 21(2), 152–159 (2005)

34. Cuff, J.A., Barton, G.J.: Application of Multiple Sequence Alignment Profiles to Improve Protein Secondary Structure Prediction. PROTEINS: Structure, Function, and Genetics 40, 502–511 (2000)

35. Cuff, J.A., Barton, G.J.: Evaluation and improvement of multiple sequence methods for protein secondary structure prediction. Proteins-Structure Function and Genetics 34(4), 508–519 (1999)

36. Wilson, C.L., Hubbard, S.J., Doig, A.J.: A critical assessment of the secondary structure α-helices and their termini in proteins. Protein Engineering Design and Selection 15(7), 545–554 (2002)

37. Rost, B., Sander, C.: Conservation and prediction of solvent accessibility in protein families. Proteins 20, 216–226 (1994)

38. Rost, B., Schneider, R., Sander, C.: Redefining the goals of protein secondary structure prediction. Journal of Molecular Biology 235, 13–26 (1994)

39. Cole, C., Barber, J.D., Barton, G.J.: The Jpred 3 secondary structure prediction server. Nucleic Acids Research 36(2), W197–W201 (2008)

40. Sander, C., Schneider, R.: Database of homology derived protein structures and the structural meaning of sequence alignment. Proteins 9, 56–68 (1991)

41. Goble, C., Stevens, R.: State of the nation in data integration for bioinformatics. Journal of Biomedical Informatics 41(5), 687–693 (2008)

RefGen: Identifying Reference Chains to Detect Topics

Laurence Longo and Amalia Todiraşcu

Abstract. In this paper, we present RefGen, the main module of a topic detection system used to improve a search engine by topic indexing. RefGen identifies reference chains and it uses genre specific properties of reference chains and (Ariel 1990)'s accessibility theory. RefGen checks several strong and weak constraints (lexical, morphosyntactic and semantic filters) to automatically identify coreference relations between referential expressions. We present the first results obtained by RefGen from a public reports corpus.

1 Introduction

We present a project aiming at automatic topic detection, by combining statistical methods and linguistic methods. We use several linguistic cues to detect topic changes: discourse markers, reference chains and theme/rheme positions. In this paper, we focus on the reference chain identification module *RefGen*, one of the main modules of our topic detection system. This topic detection system is used to improve the results of an existing search engine by using topics for document indexation.

Beside the use of explicit cue discourse phrases, we assume that topics will be mainly discovered from linguistic markers as reference chains and anaphora pairs (Cornish 1995), (Schnedecker 1997). Specific cue discourse phrases explicitly focus on a specific topic, especially at the beginning of a paragraph: *"concernant X/' concerning X'*, *premièrement..."*, *'first'*. But the cases of explicit use of discourse markers are quite rare and topics are represented by various linguistic expressions, such as reference chains. Reference chains reinforce text cohesion by pointing a single discourse entity. This discourse entity might be an actor or an event related to the current segment topic. Thus, we exploit these linguistic markers if no explicit discourse phrase has been used. These markers point out the entity frequently referred in a paragraph. First, we identify the reference chains from each discourse segment. Then, we propose local candidate topics for each document segment, selecting them from the first elements of the reference chains. The first mention refers to a new discourse entity, so it represents a potential local topic. While our goal is to select a global topic associated to the document, we

Laurence Longo · Amalia Todiraşcu
LiLPa laboratory, University of Strasbourg, 67000 Strasbourg, France

V. Pallotta, A. Soro, and E. Vargiu (Eds.): Advances in DART, SCI 361, pp. 27–40.
springerlink.com   © Springer-Verlag Berlin Heidelberg 2011

select it from these local topics, by applying selection criteria such as frequency (several chains referring to the same entity), such as position in the document, such as topic continuity among several segments (Goutsos 1997).

Many natural language processing applications such as topic detection, text summarization or human-machine dialogue systems use reference chains to build discourse model or interpretation. A reference chain includes at least three referential expressions (e.g. *"Michelle Alliot-Marie ... elle ... sa" / 'Michelle Alliot-Marie ... she ... her'*) pointing the same entity (Schnedecker 1997). To find reference chains, the systems identify the various referential expressions (e.g. pronouns, definite noun phrase, possessives) referring the same discourse entity. This entity is common to several sentences of the same paragraph and it represents a potential topic candidate.

The correct identification of the coreference relations is a difficult task. Coreference relations are identified by searching valid pairs of antecedent and anaphora candidates. The candidate pairs should verify several morpho-syntactic or semantic constraints (Grosz et al. 1995), (Beaver 2004). To find these candidate pairs, the existing systems either apply manually defined heuristic rules (which select the most suitable candidates for a given anaphor) or rules learned from annotated corpora. Coreference resolution methods require large syntactically and semantically annotated corpora or complex knowledge bases to identify coreference expressions and relations.

Even if supervised learning methods (Ng and Cardie 2002), (Hoste 2005), (Denis 2007) proved their efficiency for coreference relation identification for several languages (English, German, Dutch etc.), these methods require large, manually annotated training corpora. Few resources are available for French: some corpora annotated for anaphoric relations, or definite descriptions (Manuélian 2003), but there is no large reference corpus annotated with reference chains in French[1] (Salmon-Alt 2001) to apply machine learning techniques.

Because no French reference corpus is available, we propose a new knowledge poor method for coreference identification. We adopt robust methods such as those used for pronoun (Mitkov 2001) and for coreference resolution (Hartrumpf 2001), (Popescu-Belis 1999), (Bontcheva et al. 2002). Our approach combines a constraint-based method, based on the Optimality Theory (Beaver 2004). The *RefGen* module identifies the starting element of a reference chain and then it selects the next elements of the reference chain applying strong and weak constraints (lexical, morpho-syntactic, syntactic and semantic) (Gegg-Harrisson and Byron 2004) between antecedent-anaphor potential pairs. We select the coreference chain elements using criteria about accessibility and information content of various categories of referring expressions (Accessibility theory (Ariel 1990)) and also their syntactic function. Moreover, we exploit some genre-dependent properties of reference chains. We study several corpora and we identify these genre dependent properties. We apply these parameters to configure our system for some specific genres (public reports, newspapers).

---

[1] For example, SemEval 2010 task#1 *Coreference Resolution in Multiple Languages* campaign provides training data for different languages except French.

The paper is organised as follows. In section 2 we present the architecture of our topic detection system. In section 3 we explain the reference relation we process. In section 4 we describe the *RefGen* architecture, the genre-dependent parameters used to identify chains, the corpus analysis, the annotation scheme adopted and the reference identification algorithm (*CalcRef*). We then compare the results obtained by *RefGen* with manually annotated corpora.

2   The Architecture of the Topic Detection System

For our project, we consider that the topics are aggregates of the sentence topics (Goutsos 1997), while sentence topics are actors, ideas or events. To detect topics, we use the global properties of the text: cohesion and coherence (Halliday and Hasan 1976), but also genre-specific properties (Biber 1994). Thus, we combine statistical methods (Choi et al. 2001) and linguistic markers identification. Beside the use of explicit discourse cue phrases (Charolles 1997), we assume that topics will be mainly discovered from cohesion markers, as reference chains (Schnedecker 1997) and anaphora pairs (Kleiber 1994).

We present our topic detection system's architecture, which is still under development (Fig. 1). First, we convert the documents available in various formats (PDF, XML, etc.) to raw text. Then, we segment the documents into several topic homogeneous units, using C99 algorithm (Choi et al. 2001). C99 detects the boundaries of the topic homogeneous units using lexical-based cohesion measures, but it does not explicitly extract topics from the unit. To associate topic candidates to each segment, we apply several heuristic rules exploiting linguistic information. Thus, we use two categories of linguistic markers: explicit discourse cues, used to focus on a specific topic (Charolles 1997), (Porhiel 2004) and cohesion markers (as reference chains (Schnedecker 1997) and anaphora pairs (Kleiber 1994)). We establish a list of discourse markers (as *"concernant X" / 'concerning X'*, *"au sujet de X..." / 'About X...'*, *"dans un premier temps" / 'first'*, *"finalement" / 'eventually'*) explicitly indicating the topic of the sentence or of the paragraph. Reference chains or anaphora pairs indicate that the same entity is referred several times in the text, the introduction of a new entity and of a new reference chain is a sign of topic shift (Vonk et al. 1992). The core module of our topic detection system is *RefGen*, the reference chain identification module. This module uses several lexical, syntactic and semantic constraints described in section 4.4.

The output of the topic detection module is a set of topics associated to each document. We apply the reference chain identification algorithm to each segment, to propose some local topic candidates. We select the starting elements of the reference chains as local topics. Then, we check criteria such as frequency (several chains referring to the same entity), position in the document (if the topic occurs in the title or in the first paragraph of the document) and topic continuity among several segments, to propose topic candidates describing the document.

The topics extracted from each document are used by the search engine to index the document. Thus, it is possible to select associated documents, among the documents indexed by similar topics.

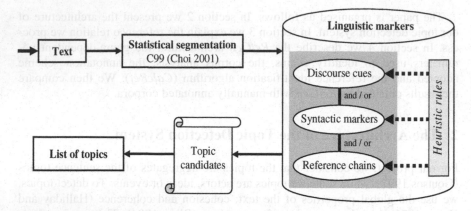

Fig. 1 The architecture of the topic detection system

3 The Reference Chains

(Schnedecker 1997) considers reference chains as cohesion markers. A reference chain contains at least three mentions (three referential expressions) referring to the same unique entity. In the example below, we have three mentions for the referent '*Mr Pons*': "*M. Pons, lui, son*"/'*Mr Pons, him, his*'

"*M. Pons rappelle que Jacques Chirac lui apparaît comme le candidat légitime de son parti*" / '*Mr. Pons said that Jacques Chirac seems to be the legitimate candidate of his political party*'. (Le Monde diplomatique).

The reference chains are composed of three types of constituents with a referential function: the proper nouns, the noun phrases (definite, indefinite, possessive or demonstrative) and the pronouns. The proper nouns play an important role in the discourse structure because they are often opening a reference chain in the journalistic portraits (Schnedecker 2005). Beside the cases of referential competition (in this case, the repetition of the proper noun eliminates ambiguity between two referents), the repetition of a proper noun signals a break in the reference chain. Each referring expression triggers a "particular recruitment process" of a referent. Thus, the demonstrative (e.g. "*ce commissaire*" / '*this commissioner*') points directly to the referent on the basis of proximity while the anaphoric pronoun "*il*" recruits a referent that is the argument of a salient phrase (Kleiber 1994). The use of a particular mention (referential expression) is an indication for the reader to remember a specific referent and which is a local theme. However, the use of noun phrases in contexts that a pronoun would be sufficient is an indication of the end of a reference chain. These elements will be used by the topic detection system.

We process single referential relations between co-referent expressions, excluding plural anaphora ("*Barack et Michèle Obama … le couple presidential … Michèle*" / '*Barack and Michele Obama … the presidential couple … Michele*'). We treat direct coreference (Manuélian 2002) for the coreferential noun phrases (NPs) sharing the same lexical head (eg "*les effets du changement climatique / ces effets"*; *'the effects of climate change / these effects'*) and some indirect coreference between person and function name (e.g. *"Barack Obama ... le président des Etats-Unis"* / *'Barack Obama ... president of the United States'*). In the future extensions of the system, we will also treat hyponym and hyperonym coreference cases.

4 The RefGen Module

According to (Schnedecker, 2005), we assume that reference chains have specific linguistic properties depending on the text genre (explanatory, narrative, argumentative etc.) and we exploit these properties for reference chains identification.

In the section below, we present the study of reference chains properties on a corpus of several genres. Then, we focus on the *RefGen* architecture and on the linguistic annotations required (tagging, chunking and automatic named entities recognition). We explain the reference chains algorithm (*CalcRef*) and we discuss the results of the evaluation.

4.1 *Corpus Analysis*

To identify reference chains properties we study the reference chains in a French corpus (about 50 000 tokens) composed of five various genres (Longo and Todirascu 2010):

- newspapers from *Le Monde* (2004),
- editorials from *Le Monde Diplomatique* (1980-1988),
- a novel *Les trois Mousquetaires* (Dumas, 1884),
- some European legal standards from the *Acquis Communautaire* (Steinberger et al. 2006)
- public reports from *La Documentation Française* (2001).

We manually annotate the reference chains to determine which reference chain properties are relevant for a particular genre. The corpus is quite small, due to the difficulties of annotation process.

For each genre we examine the chains following five criteria (Schnedecker 2005):

- the number of mentions (the average length of chains);
- the average distance between the mentions (the number of sentences);
- the grammatical class of the mentions and their frequency;
- the grammatical class of the first mention;
- the identity between the sentence theme and the first mention of a chain.

The study revealed several differences across genres. For example, reference chains from newspapers are relatively short (the average length is four) while we find long chains for the novel (the length is nine). Concerning the frequency of the referential categories, we notice that the newspapers contain mostly proper nouns

(30.8 %) while editorials contain 50 % of definite NP. Definite NPs are very frequent first mentions for public reports while proper nouns are starting elements of reference chains for newspapers. In addition, the first element of the chain is the sentence topic for 80 % of the occurrences for the newspapers and only for 40 % for the public reports. This last criterion is used to know if we could gather the reference chains containing the same sentence topic (coreferent coreference chains (Schnedecker 1997)) to identify the document topic.

Thus, the corpus analysis on the reference chains highlights their genre-specific properties (see Fig 2.), used to configure *RefGen* according to the genre.

corpus / criteria	Newspapers	Editorials	Laws	Novel	Public reports
Length of chain	4	3,7	3	9	3,4
Distance between mentions	0,8	0,9	0,6	0,4	1,1
Grammatical class of the 1st mention	Proper noun	Complete NP	Indefinite NP	Indefinite NP	Definite NP
F of mentions	30% proper noun	50% definite NP	40% indefinite NP	36% pronoun	- 33% pronoun - 33% definite NP
Identity theme -1st mention	80%	100%	60%	60%	40%

Fig. 2 The genres and their properties

4.2 Architecture of RefGen

RefGen is composed of several modules (Fig 3). The raw input text is tagged, lemmatized and annotated at chunks level with TTL (Ion, 2007). In addition, we apply several annotation modules to automatically identify the referential expressions (named entities, complex noun phrases) that may be potential candidates as the first mention of a reference chain. Also, we automatically annotate the impersonal occurrences of the French pronoun "*il*" to avoid wrong anaphora candidates (the system will ignore these impersonal non-anaphoric occurrences).

Using these automatic annotations, *CalcRef* computes the reference chains. It first selects the potential candidates for the first mention of a chain (according to

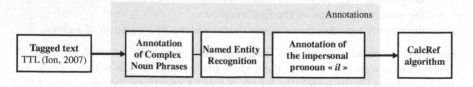

Annotations

Tagged text TTL (Ion, 2007) → Annotation of Complex Noun Phrases → Named Entity Recognition → Annotation of the impersonal pronoun « *il* » → CalcRef algorithm

Fig. 3 The *RefGen* modules

their global accessibility weight, their syntactic function and the genre specific parameters). Then, it selects the next elements of the chain from antecedent-anaphora pairs which satisfy several lexical, syntactic and semantic constraints.

4.3 Automatic Annotation of the Referential Expressions

We tagged, lemmatized and chunked the documents using TTL tagger (Ion 2007). This tagger uses the MULTEXT tagset (Ide and Véronis 1994) and XML output. It identifies lemmas, chunks (simple noun phrases, prepositional phrases) and some morpho-syntactic properties (tense, mode, person, gender, number).

For example, TTL proposes the following output for *"le ministre des affaires étrangères"* / *'the Foreign Affairs Minister'* (Fig 4).

```
<s id="ttlfr.3">
   <w lemma="le" ana="Da-ms" chunk="Np#1">Le</w>
   <w lemma="ministre" ana="Ncms" chunk="Np#1">ministre</w>
   <w lemma="de_le" ana="Dg-fp" chunk="Pp#1,Np#2">des</w>
   <w lemma="affaire" ana="Ncfp" chunk="Pp#1,Np#2">affaires</w>
   <w lemma="étranger" ana="Af-fp" chunk="Pp#1,Np#2,Ap#1">étrangères</w>
```

Fig. 4 Example of TTL (Ion 2007) output

In this example, we have a noun phrase (Np#1) *"le ministre"* and a preposi-tional phrase *"des affaires étrangères"* (Pp#1). These informations are used to automatically identify the referential expressions.

Then, we apply a set of rules to automatically identify complex noun phrase. A complex noun phrase (CNp) is a Np modified by at most two Pps or a Np modi-fied by a relative clause; as *"l'utilisation des fonds publics"* / *'the use of the public funds'* which are more informative than simple Np *"l'utilisation"* / *'the use'*. This step is required to select possible first mentions of a reference chain from all the referential expressions. To identify the CNp, we create 122 morpho-syntactic pat-terns. In the example below, the Np *"l'utilisation"* and the Pp *"des fonds publics"* are gathered in a single CNp (CNP#1):

```
<s id="ttlfr.6">
   <w lemma="le" chunk="CNP#1, Np#1" pattern="85" ana="Da-fs">L'</w>
   <w lemma="utilisation" chunk="CNP#1, Np#1" pattern="85" ana="Ncfs">utilisation</w>
   <w lemma="de_le" chunk="CNP#1, Pp#1, Np#2" pattern="85" ana="Dg-mp">des</w>
   <w lemma="fond" chunk="CNP#1, Pp#1, Np#2" pattern="85" ana="Ncmp">fonds</w>
   <w lemma="public" chunk="CNP#1, Pp#1, Np#2, Ap#1" pattern="85" ana="Af-mp">publics</w>
```

Fig. 5 Example with the CNp annotations

After CNp identification we apply the Named Entity Recognizer module. This module applies some heuristic rules (140 rules) to automatically identify person names (Pers) and organization names (Org). We also automatically annotate the function names (func) (*"président"*/*'president'*, *"directeur"*/ *'director'*,

"*colonel*"/'*colonel*') for two reasons. First, because function names may be associated with a person name (relation between the person name and its function), as
"*Dominique Muller – président du conseil*" / '*Dominique Muller – chairman of
the board*' to reinforce the link between entities. Second, because function names
are good proofs to categorize person names and organisation names (internal and
external evidences (McDonald 1996)). In this sense, our named entity extractor
first defines the boundaries of the named entity (by adding the attribute
'Ner="NER"') and then it assigns the type of the entity ("Org", "Pers",
"Func"). For example, we annotate "*France 2*" as an organization name
("NER#5, Org") with help to the number "2" which is an internal evidence
(Fig. 6):

```
<w lemma="sur" chunk="Pp#2" pattern="60" ana="Sp">sur</w>
<w lemma="France" chunk="Pp#2, Np#6" pattern="17, 60, 12" Ner="NER#5, Org" ana="Np">France</w>
<w lemma="2" chunk="Pp#2, Np#6" pattern="17, 60, 12" Ner="NER#5, Org" ana="Mc">2</w>
```

Fig. 6 Example with the named entity annotations

In addition, we automatically annotate the French impersonal pronoun *il* (e.g.
"*il* pleut", '*it* rains') to eliminate the non-anaphoric use of this pronoun (Fig 7).
We use some lexical lists (weather verbs, past participles predominantly impersonal and adjectives) to create 382 morpho-syntactic patterns to automatically
identify the impersonal use of the pronoun "*il*". The pronouns annotated
'feat="imp"' will be ignored by *CalcRef* (the reference chain algorithm) when
checking possible anaphor. In the following example, the use of the pronoun "*il*"
in "*il importe de veiller à*"/'*it is important to ensure*' is impersonal, so it is automatically tagged 'feat="imp"' and it is ignored by *CalcRef*:

```
<w lemma="il" chunk="Vp#2" pattern="376" ana="Pp3ms" feat="imp">il</w>
<w lemma="importer" chunk="Vp#2" pattern="376" ana="Vmsp3s">importe</w>
<w lemma="de" chunk="Vp#3" pattern="376" ana="Spd">de</w>
<w lemma="veiller" chunk="Vp#3" pattern="376" ana="Vmn">veiller</w>
<w lemma="à" ana="Spa">à</w>
```

Fig. 7 Example with the impersonal pronoun annotation

We use these automatic linguistic annotations to identify the reference chains
and anaphora pairs.

4.4 The CalcRef Module

The core module of *RefGen, CalcRef,* identifies reference chains from the annotated texts. To do this, *CalcRef* uses genre-dependent parameters and the linguistic
annotations presented in the previous section. While the topic search engine is
genre-dependent, we configure genre-specific parameters to match the genre of

the indexed documents. Genre-based properties of the reference chains were presented in section 4.1.: the average distance between the mentions, the average chain length, the preferred category of the first element of a chain. Thus, for a corpus of public reports, we use the length of 4, the average distance is 2 phrases and the preferred type of the first element is a complete definite noun phrase. For a newspaper corpus, the length is 3 and the distance is 1.

For each text segment (obtained after statistical topic segmentation (Choi et al. 2001)), *CalcRef* searches candidates for the first element of reference chains, from the expressions with a high degree of information content. (Ariel 1990) classifies these referential expressions according to their accessibility in discourse and in speaker's mind. The accessibility is computed by combining three elements: informativity (the amount of lexical information), rigidity (the possibility to pick up a specific referent) and attenuation (phonological size). Less the referent is accessible, it is long, rigid and contains rich information about the referred entity. Accessibility weights are computed as sum of rigidity, of attenuation and of informativity. Each element is computed from 10 to 110. For example, the global weight of the complete proper noun *"Le PDG Marcel Klauss"*/' *CEO Marcel Klauss'* is 220 while it is 150 for the pronoun *"il"*.

As first mention candidates, we select less accessible elements: proper nouns, CNp (definite Np modified by Pps or by a relative clause) (Fig 8). Indefinite Nps were added to the initial Ariel's hierarchy, due to the fact that indefinites might also start reference chains (for explanatory or descriptive texts). Short mentions[2] as pronouns are used to refer to entities already specified in the discourse.

To select the first mentions of a reference chain or a possible antecedent, *CalcRef* uses the global weight of the candidates. The global weight is the sum of the global accessibility weight and the syntactic role weight. Thus, each expression weight is augmented by a syntactic weight. The syntactic weight is 100 for the subject, 50 for the direct object, 30 for the indirect object and 20 for other syntactic functions. Then, we use genre-dependent parameters (such as the preference for the first element type or the distance between the mentions) to increase the weight (+50) of some candidates. For example, if we treat newspapers, proper nouns are frequently first mentions of a chain, while definite Nps are preferred as first elements of a chain for public reports. The first mention candidates are ordered according to their global weight. The highest weight candidates are selected as first mention of possible chains.

Anaphora candidates are selected by *CalcRef* from highly accessible expressions (pronouns, demonstratives etc.). Then, the module selects possible antecedents from low accessible expressions, such as complete proper nouns or definite Nps, found in the context of the anaphora candidates, in a window of n words. The value of n should be less than the average distance between the mentions. Potential antecedent-anaphor pairs should satisfy several morpho-syntactic and semantic properties (Beaver 2004).

---

[2] Mentions means also referential expression.

Mention	Informativity	Rigidity	Attenuation	Global Accessibility
Indefinite noun group	110	110	10	230
Complete modified Proper Noun	100	100	20	220
Proper Noun	90	90	30	210
Complex definite noun phrase	80	80	40	200
Simple definite noun phrase	70	70	50	190
Last name	60	60	60	180
First name	50	50	70	170
Demonstrative	40	40	80	160
Pronoun	30	30	90	150
Reflexive pronoun	20	20	100	140
Possessive	10	10	110	120

Fig. 8 Accessibility table

We adapt the method proposed by (Gegg-Harrison and Byron 2004), by defining several constraints between antecedent and anaphor to filter out impossible pairs. For each pair, we check some strong and weak constraints, defined for French.

Weak constraints mean that they might not be satisfied, even if there is a valid antecedent-anaphor pair: agreement in gender or number, similar syntactic function, semantic knowledge (person names might be valid antecedents of a noun phrase expressing a function). For example, we might have a different syntactic function between antecedent ("*le rapport annuel concernant l'emploi de la langue française*"/'*annual report on the use of French*') and anaphor ("*ce rapport*"/'*this report*'), and the two elements are related by a coreference relation:

"*La Commission a présenté au Parlement **le rapport annuel concernant l'emploi de la langue française**. **Ce rapport** a tiré un signal d'alarme ...*"

'*The Commission presented in the Parliament the annual report concerning the use of French language. This report sent an alarm message...*'

Strong constraints should not be violated for valid candidate pair. Indeed if an element is nested in its antecedent as *[les mesures [contre les émissions de gaz à effet de serre]]*), or if the two elements are co-arguments of a verb, then the pair is not valid.

For each candidate pair satisfying the strong constraints, we check the number of the weak constraints that are satisfied. In the case of the several pairs satisfying the same number of constrains for the same anaphor, we keep the valid pairs into a large list.

For each paragraph, *CalcRef* establishes a list of candidates of first mentions and a list of possible antecedent-anaphor pairs. To build the reference chain from the set of valid antecedent-anaphor pairs, *CalcRef* selects a first element candidate from the current list and it searches the pairs having this candidate as antecedent. To build the reference chain, *CalcRef* assumes that coreference relation is transitive: if A is antecedent of B and B is antecedent of C, then they are part of the same chain. The process of searching coreference chains continues until the length of the current reference chain is greater than the average gender-specific length. The candidate pairs identified as part of the current reference chain are annotated (with the "`coref`" attribute) (Fig 9).

```
<segments>
- <seg lang="fr">
  - <s id="ttlfr.1">
      <w lemma="le" chunk="Np#1" ana="Da-fs" coref="1">La</w>
      <w lemma="décision" chunk="Np#1" ana="Ncfs" coref="1">décision</w>
      <w lemma="aller" chunk="Vp#1" ana="Vmip3s">va</w>
      <w lemma="t" ana="U">-t</w>
      <w lemma="il" ana="Pp3fs" coref="1">-elle</w>
      <w lemma="forcer" ana="Vmn">forcer</w>
      <w lemma="Google" chunk="Np#2" pattern="17, 60, 97" ner="NER#1,
      Org" ana="Np" coref="2">Google</w>
      <w lemma="à" chunk="Vp#2" ana="Spa">à</w>
      <w lemma="infléchir" chunk="Vp#2" ana="Vmn">infléchir</w>
      <w lemma="son" chunk="Np#3" ana="Ds3fp" coref="2">ses</w>
      <w lemma="pratique" chunk="Np#3" ana="Ncfp">pratiques</w>
      <c>?</c>
  </s>
</seg>
```

Fig. 9 Annotation of a candidate pair

The whole process is restarted for each potential first mention of a reference chain and for each paragraph.

4.5 *Evaluation*

We compare the reference chains extracted automatically by *RefGen* against a manually annotated corpus. The evaluation corpus is quite small (7230 tokens). It is composed of public reports of the European Commission about the climate changes and the measures adopted by Europe to limit the effects of the climate changes. We define our own manually annotated corpus, because no French reference corpus is available. Even if we annotate only the coreference relations that we are able to process, the annotation process is long and difficult for humans.

We present the results obtained for the CNp annotation module, for the named entity recognizer (NER) module and for the reference chain identification module. We compute the recall, the precision and the f-measure of the intermediate modules, as well as the results for *CalcRef*. We check the results obtained for independent antecedent – anaphor pairs, as well as for reference chains. We evaluate *CalcRef* using two different configurations (genre-dependent parameters): for public reports and for newspapers (Fig 10). The genre-dependent parameters for public reports are the following: the average distance is 2, the average length is 4 and the preferred type is definite description. The second configuration (specific to newspapers) consists of: the distance = 1, the length = 3 and the preferred type = proper noun.

	NER	CNp	CalcRef (pairs)	CalcRef (reference chains)
recall	0,85	0,87	0,69	0,58
precision	0,91	0,91	0,78	0,70
f-measure (public reports)	0,88	0,89	0,73	0,63
f-measure (newspapers)	0,88	0,89	0,70	0,54

Fig. 10 The evaluation of *RefGen's* modules : NER module, CNp module and *CalcRef* module

Thus, the NER module fails to identify some acronyms or abbreviations (e.g. *GES: "gaz à effet de serre"* / *'greenhouse gas emissions'*) which were annotated as Organizations. The CNp identification module does not identify some CNps because the existing set of patterns does not contain an appropriate pattern. NER and CNp annotation errors have as effect that some mentions of the reference chain are not correctly identified.

For the public reports corpus with its genre-dependent parameters, *CalcRef* identifies 118 pairs, but only 24 could be related by reference chains. For evaluation, we count the number of common links (antecedent-anaphor pairs) between system reference chains and manually annotated reference chains. Several antecedent-anaphor pairs were wrongly selected, due to tagging errors or due to the insufficient external knowledge sources. For example, some of the antecedent-anaphor pairs were selected because they satisfy the same number of constraints (number, gender, syntactic function) (*"ils– les efforts d'adaptation entrepris par les États membres"* / *'they – adaptative efforts of Member States'* and *"ils– des Etats membres"* / *'they – Member States'*). For the two configurations (with public reports genre-dependent parameters and newspapers genre-dependent parameters), we notice a small variation of the f-measure for pair identifications (0,70 in the newspapers configuration instead of 0,73 for public reports). The f-measure for the reference chain identification with the newspapers parameters is worse (0,54) than public reports configuration (0,63).

5 Conclusion

We presented *RefGen*, a knowledge poor, robust reference chain identification module, developed for French. This module uses a set of detailed linguistic annotation and the accessibility hierarchy of the referring expressions to select possible mentions and antecedent-anaphor candidates. A set of lexical, syntactic and semantic constraints are used to filter some pairs. *RefGen* also uses some genre-dependent properties of the reference chains (average length, preferred type of the first element, average distance separating several mentions of the same referent).

These genre-dependent properties were identified from a corpus-based analysis. We describe the algorithm adopted to identify the reference chains and we present a first evaluation of the module. A first evaluation of *RefGen* shows that the results are influenced by the genre-specific parameters. We will evaluate the module on a larger test corpus and for other genres.

In the future, the module will be integrated into the topic detection system. Future work focus on the adaptation of the system for other languages.

References

Ariel, M.: Accessing Noun-Phrase Antecedents. Routledge, London (1990)

Beaver: The optimization of discourse anaphora. Linguistics and Philosophy 27(1), 3–56 (2004)

Biber, D.: Representativeness in corpus design. Linguistica Computazionale, IX–X, Current Issues in Computational Linguistics: in honor of Don Walker (1994)

Bontcheva, K., Dimitrov, M., Maynard, D., Tablan, V., Cunningham, H.: Shallow methods for named entity coreference resolution. In: Proceedings of TALN 2002 (2002)

Charolles, M.: L'encadrement du discours: univers, champs, domaines et espaces, Cahier de Recherche Linguistique 6, LANDISCO, Université Nancy 2, 1–73 (1997)

Choi, F.Y.Y., Wiemer-Hastings, P., Moore, J.: Latent semantic analysis for text segmentation. In: Proceedings of NAACL 2001, pp. 109–117 (2001)

Cornish, F.: Références anaphoriques, références déictiques, et contexte prédicatif et énonciatif. Sémiotiques 8, 31–57 (1995)

Denis, P.: New Learning Models for Robust Reference Resolution. PhD thesis, University of Texas, Austin (2007)

Gegg-Harrison, W., Byron, D.: PYCOT: An Optimality Theory-based Pronoun Resolution Toolkit. In: Proceedings of LREC 2004, Lisbonne (2004)

Goutsos, D.: Modeling Discourse Topic: sequential relations and strategies in expository text. Ablex Publishing Corporation, Norwood (1997)

Grosz, B.J., Weinstein, S., Joshi, A.K.: Centering: a framework for modeling the local coherence of discourse. Computational Linguistics 21(2), 203–225 (1995)

Halliday, M., Hasan, R.: Cohesion in English. Longman English Language Series, vol. 9. Longman, London (1976)

Hartrumpf, S.: Coreference Resolution with Syntactico-Semantic Rules and Corpus Statistics. In: Proceedings of CoNLL (Computational Natural Language Learning Workshop) (2001)

Hoste, V.: Optimization Issues in Machine Learning of Coreference Resolution. PHD thesis, p. 246 (2005)

Ide, N., Veronis, J.: MULTEXT (Multilingual Tools and Corpora). In: Proceedings of the 14th International Conference on Computational Linguistics, Kyoto (1994)

Ion, R.: TTL: A portable framework for tokenization, tagging and lemmatization of large corpora. Romanian Academy, Bucharest (2007)

Kleiber, G.: Anaphores et Pronoms. Duculot, Louvain-la-Neuve (1994)

Longo, L., Todirascu, A.: Une étude de corpus pour la détection automatique de thèmes. In: Proceedings of the 6th Journées de Linguistique de Corpus (JLC 2009), Lorient, France (2010)

Manuélian, H.: Annotation des descriptions définies: le cas des reprises par les rôles thématiques. In: Proceedings of RECITAL 2002, Nancy, France, pp. 455–467 (2002)

Manuélian, H.: Descriptions définies et démonstratives: analyse de corpus pour la génération de textes. PhD thesis, Université de Nancy 2, France (2003)

Mitkov, R.: Towards a more consistent and comprehensive evaluation of anaphora resolution algorithms and systems. Applied Artificial Intelligence: An International Journal 15, 253–276 (2001)

Ng, V., Cardie, C.: Improving machine learning approaches to coreference resolution. In: Proceedings of the ACL (Association For Computational Linguistics), Morristown, pp. 104–111 (2002)

Popescu-Belis, A.: Modélisation multi-agent des échanges langagiers: application au problème de la référence et à son évaluation. PhD thesis, Université Paris-XI (1999)

Porhiel, S.: Les introducteurs thématiques. Cahiers de Lexicologie 85 (2004)

Salmon-Alt, S.: Référence et Dialogue finalisé: de la linguistique à un modèle opérationnel. PhD thesis, Université H. Poincaré, Nancy (2001)

Schnedecker, C.: Nom propre et chaînes de référence. Recherches Linguistiques, vol. 21. Klincksieck, Paris (1997)

Schnedecker, C.: Les chaînes de référence dans les portraits journalistiques: éléments de description. Travaux de Linguistique 51, 85–133 (2005)

Vonk, W., Hustinx, L., Simons, W.: The use of referential expressions in structuring discourse. Language and Cognitive Processes 7, 301–333 (1992)

Synonym Acquisition across Domains and Languages

Lonneke van der Plas, Jörg Tiedemann, and Jean-Luc Manguin

Abstract. We describe an approach to the automatic extraction of synonyms that
is easy to port across domains and across languages. The approach relies on automatic word alignments in parallel texts and uses distributional methods to compute
the semantic similarity of words based on these word alignments. As a result the
system outputs ranked lists of candidate synonyms for a given word. We apply the
method to French, a language for which an extensive electronic synonym dictionary
is available, that serves to evaluate the method. We compare the performance with
a system that uses syntactic contexts to acquire synonyms automatically. We show
that the alignment-based method outperforms the syntactic method by a large margin. In addition, we show that we can adapt to the domain of colloquial language
use by replacing the parallel corpus with one that contains a lot of conversational
speech: a corpus of movie subtitles. Furthermore, we apply the method to another
language, Dutch, with similar performances.

1 Introduction

Support for semantics has been mentioned as one of the goals of next generation
information retrieval tools. Synonymy is a type of lexico-semantic knowledge that
helps to overcome the so-called terminological gap for tasks such as information
retrieval, information extraction, and question answering. Imagine a French student

Lonneke van der Plas
University of Geneva, Switzerland
e-mail: lonneke.vanderplas@unige.ch

Jörg Tiedeman
Uppsala University, Sweden
e-mail: jorg.tiedemann@lingfil.uu.se

Jean-Luc Manguin
CNRS/University of Caen, France
e-mail: jean-luc.manguin@unicaen.fr

V. Pallotta, A. Soro, and E. Vargiu (Eds.): Advances in DART, SCI 361, pp. 41–57.
springerlink.com   © Springer-Verlag Berlin Heidelberg 2011

searching for a job types *cherche boulot* into a search engine. The student might be ignorant to the fact that the word *boulot* is a colloquial term for synonyms such as *travail, poste*, while these latter terms (*travail, poste*) are used in the majority of job announcements. Hence, simple word matching will not retrieve the information needed by the student. Resources that group French synonyms could help this user in fulfilling his/her information need.

Synonym dictionaries are a common source of semantic information that could be used to deal with the problem described above. However, the drawback of dictionaries is that they are static and based on common knowledge, and therefore not personalised. Personalisation and context awareness have been mentioned as goals to improve information retrieval tools in addition to support for semantics. Providing domain-dependent lexical information is a first step towards context-aware search engines. Search engines need to know when to relate a word like *bank* with the establishment for the custody of money (in the financial domain, for example) and when to relate it to the shore of a river. There are domain-specific dictionaries available, but the number of domains covered is limited.

Automatic methods for synonym acquisition are more flexible and therefore more easily adjustable to emerging needs. For example, work on the acquisition of synonyms using distributional models has shown that syntactic contexts can be applied to any large corpus of text that is analysed syntactically to acquire semantically related words [12, 15] and the method has been applied to corpora from different domains [21] with reasonable success. However, one of the prerequisites for this method is a large parsed corpus or at least a syntactic parser for the target language. For English there are many parsers available but for the majority of languages such tools do not exist. The syntax-based method for the acquisition of synonyms cannot be applied to those languages. A personalised search tool would at least want to serve the user in his/her own language. Therefore, we need automatic methods for synonym acquisition that are easily portable across different languages and that rely as little as possible on language-specific pre-processing.

In this paper we will present a method that is particularly well-suited to be ported across different languages and across different domains. Moreover, the method outperforms the syntax-based approach described above for the task of synonym acquisition.

2   The Distributional Hypothesis

Before we move to describing the methodology we need to explain the hypothesis that underlies our work, the distributional hypothesis. It states that semantically related words are distributed similarly over contexts [11]. In other words, you can grasp the meaning of a word by looking at its contexts.

Context can be defined in many ways. Previous work has been mainly concerned with the syntactic contexts a word is found in. For example, the verbs that are in a subject relation with a particular noun form a part of its context. These contexts can be used to determine the semantic relatedness of words. For instance, words that occur in a object relation with the verb *to drink* have something in common: they are liquid.

With the advent of multilingual parallel corpora, yet another type of context has been born, the multilingual, translational context. Moreover, tools from the machine translation community such as automatic word alignment tools, that we will discuss in more detail in the next section, have opened the way to rather precise multilingual contexts on the word level. We can extract a list of probable translations for a word in the languages included in the parallel corpus from the output of word-alignment tools. In addition, we can compute the number of times the tool has found a given translation pair. The translational context of a word is composed of the set of translations it gets in other languages. For example, the translational context of *cat* is *kat* in Dutch and *chat* in French. How do we get from translational contexts to synonymy? The idea is that words that share a large number of translations are similar. For example both *autumn* and *fall* get the translation *herfst* in Dutch, *Herbst* in German, and *automne* in French. This indicates that *autumn* and *fall* are synonyms.

Bilingual dictionaries are another source of translations of words. Although dictionary information is very precise and less noisy than the translational contexts automatically acquired from multilingual parallel corpora, there are several advantages associated with automatically extracted translational contexts. Dictionaries are not always publicly available for all languages. Dictionaries are static and often incomplete resources, and they do not provide frequency information. For the acquisition of translational contexts, any multilingual parallel corpus can be used. It is thus possible to focus on a special domain. Furthermore, the automatic alignment provides us with frequency information for every translation pair, useful for handling ambiguity.

3   Translational Context

We rely on automatic word alignment in parallel corpora to find the most probable translation pairs.

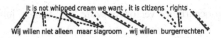

Fig. 1 Example of bidirectional word alignments of two parallel sentences

Figure 1 illustrates the automatic word alignment between a Dutch and an English phrase as a result of using the IBM alignment models [4] implemented in the open-source tool GIZA++ [19]. The alignment of two texts is bidirectional. The Dutch text is aligned to the English text and vice versa (dotted lines versus continuous lines). The alignment models produced are asymmetric. Several heuristics exist to combine directional word alignments. The intersection heuristic, for example, only accepts translation pairs that are found in both directions.

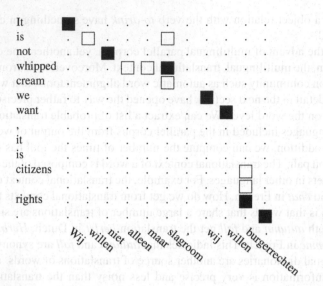

Fig. 2 A word alignment matrix

In Figure 2 we see the intersection of links illustrated by filled boxes. Additional alignment points from the union of links are shown as empty boxes.

Translational contexts are used to find distributionally similar words. We give some examples of translational co-occurrence vectors in Table 1. Every cell in the vector refers to a particular translational co-occurrence type. For example, *chat* 'cat' gets the translation *Katze* in German. The value of these cells indicate the number of times the co-occurrence type under consideration is found in the corpus.

Each co-occurrence type has a cell frequency. Likewise each head term has a row frequency. The row frequency of a certain head term is the sum of all its cell frequencies. In our example the row frequency for the term *chat* 'cat' is 60. Cut-offs for cell and row frequency can be applied to discard certain infrequent co-occurrence types or head terms respectively.

Table 1 Translational co-occurrence vector for *poste* 'job' *boulot* 'job', and *chat* 'cat' based on four languages

	Arbeit-DE	baan-NL	lavoro-IT	job-EN	cat-EN	Katze-DE
poste	17	26	8	13	0	0
boulot	6	12	7	10	0	0
chat	0	0	0	0	26	34

For comparison, an example of a syntax-based co-occurrence vector is given in Table 2.

Table 2 Syntactic co-occurrence vector for *chat*

	avoir_obj 'have_obj'	nourrir_obj 'feed_obj'	noir_adj 'black_adj'	pliant_adj 'folding_adj'
kat 'cat'	50	10	5	1

3.1 Measures for Computing Similarity

The more similar the vectors are, the more distributionally similar the head terms are. We need a way to compare the vectors for any two head terms to be able to express the similarity between them by means of a score. Various measures can be used to compute the distributional similarity between terms. We will explain in section 4 what measures we have chosen in the current experiments.

Furthermore, it has been shown that distributional methods benefit from using feature weights. For example in syntax-based approaches selectionally weak [27] or *light* verbs such as *hebben* 'to have' are given a lower weight than a verb such as *uitpersen* 'squeeze' that occurs less frequently. We have used weights for the translational context to counter balance the alignment errors that often occur with frequent words.

3.2 Related Work

Multilingual parallel corpora have been used for tasks related to word sense disambiguation such as target word selection [9] and separation of senses [28, 10, 13].

Automatic acquisition of paraphrases using multilingual corpora is discussed in [30, 2, 5], of which only the last two are based on automatic word alignment.

[2] use a method that is rooted in phrase-based statistical machine translation. Translation probabilities provide a ranking of candidate paraphrases. These are refined by taking contextual information into account in the form of a language model. The Europarl corpus [14] is used. A precision of 55.3% is reached when using context information. A precision score of 55% is attained when using multilingual data. Manual alignment improves the performance by 26%. In a more recent publication, [5] improved this method by using syntactic constraints and multiple languages in parallel.

Improving the syntax-based approach for synonym identification using bilingual dictionaries and parallel corpora has been discussed in [16], [34], [23], and [25].

[16] try to tackle the problem of identifying synonyms in lists of nearest neighbours in two ways: Firstly, they look at the overlap in translations of semantically similar words in multiple bilingual dictionaries. Secondly, they design specific patterns designed to filter out antonyms. They evaluate on a set of 80 synonyms and 80 antonyms from a thesaurus that are also found among the top-50 distributionally similar words of each other. The pattern-based method results in a precision of 86.4 and a recall of 95.0. The method using bilingual dictionaries gets a higher precision score (93.9). However, recall is much lower: 39.2.

[34] report an experiment on synonym extraction using bilingual resources (an English-Chinese dictionary and corpus) as well as monolingual resources (an English dictionary and corpus). Their monolingual corpus-based approach is very similar to our monolingual corpus-based approach. The bilingual approach is different from ours in several aspects. Firstly, they do not take the corpus as the starting point to retrieve word alignments. They use the bilingual dictionary to retrieve multiple translations for each target word. The corpus is only employed to assign probabilities to the translations found in the dictionary. The authors praise the method for being able to find synonyms that are not in the corpus as long as they are found in the dictionary. However, the drawback is that the synonyms are limited to the coverage of the dictionary. The aim of automatic methods in general is precisely to overcome the limited coverage of such resources. A second difference with our system is the use of a bilingual parallel corpus whereas we use a multilingual corpus containing 11 languages in total. The authors show that the bilingual method outperforms the monolingual methods both in recall and precision. However, a combination of different methods leads to the best performance. A precision of 27.1 on middle-frequency nouns is attained.

In [23] the distributional alignment-based method is introduced. A comparison is made between the alignment-based method and the syntax-based method for Dutch synonym acquisition. The alignment-based method outperforms the syntax-based method. The setup in section 8 of this article does not follow [23], but experiments undertaken in [22], because the evaluation framework in [22] is more carefully designed. The latter includes a subsubsection (Section 4.5.4) on comparing different corpora for the acquisition of synonyms for Dutch.

In [25] the syntax-based and alignment-based method for French synonym acquisition on the general domain are compared. The results in Section 6 are taken from this study.

4 Materials and Methods

In the following subsections we describe the data collection we used and the similarity measure and weighting function we chose.

4.1 Data Collection

We need a parallel corpus of reasonable size with French either as source or as target language. Furthermore, we would like to experiment with various languages aligned to French. The freely available Europarl corpus [14] includes 11 languages in parallel, it is sentence aligned [32], and it is of reasonable size. Each language contains around 1 million sentences, which corresponds roughly to 28 million words. Thus, for acquiring French synonyms we have 10 language pairs with French as source language: Danish (DA), German (DE), Greek (EL), English (EN), Spanish (ES), Finnish (FI), Dutch (NL), Italian (IT), Portuguese (PT), and Swedish (SV). We applied a lemmatiser [29] to the French part of the language pairs in order to 1) reduce data sparseness, and 2) to facilitate our evaluation based on comparing our results to existing synonym databases.

Context vectors are populated with the links to words in other languages extracted from automatic word alignments. We applied GIZA++ and the intersection heuristic as explained in section 3. From the word-aligned corpora we extracted translational co-occurrence types, pairs of source and target words in a particular language with their alignment frequency attached. Each aligned target word is a feature in the (translational) context of the source word under consideration. We removed word type links that include non-alphabetic characters to focus our investigations on real words and we transformed all characters to lower case.

In Section 3, we explained that each headword has a corresponding row frequency, and each cooccurrence type has a cell frequency. Cut-offs for cell and row frequency can be applied to discard certain infrequent and often less reliable co-occurrence types or head terms respectively. We applied a cell and row frequency cutoff of 4 and 10 respectively, because these settings performed well in previous work [25].

Note that we rely entirely on automatic processing of our data. Thus, the results from automatic tagging, lemmatisation and word alignment include errors.

4.2 Comparing Vectors

To estimate the similarity of words by means of their associated translational co-occurrence vectors we need a similarity measure. We explained in 3.1 that some attributes contain more information than other attributes. We want to account for that using a weighting function, that will modify the cell values.

We have limited our experiments to using Dice†, a variant of Dice as our similarity measure and and Pointwise mutual information (MI, [6]) as weight. Dice†[1] is defined as:

$$Dice\dagger = \frac{2\sum min(weight(W1, *_r, *_{w'}), weight(W2, *_r, *_{w'}))}{\sum weight(W1, *_r, *_{w'}) + weight(W2, *_r, *_{w'})}$$

We describe the functions using an extension of the notation used by [15], adapted by [7]. Co-occurrence data is described as tuples: $\langle word, language, word' \rangle$, for example, $\langle chat, EN, cat \rangle$.

Asterisks indicate a set of values ranging over all existing values of that component of the relation tuple. For example, $(w, *, *)$ denotes for a given word w all translational contexts it has been found in any language. For the example of chat, this would denote all values for all translational contexts the word is found in: Katze_DE:17, chat_FR:26 etc. There is a placeholder for the weighting function: weight.

Pointwise mutual information (MI) measures the amount of information one variable contains about the other. MI is computed as follows:

---

[1] Note that Dice † gives the same ranking as the well-known Jaccard measure, i.e. there is a monotonic transformation between their scores. Dice † is easier to compute and therefore the preferred measure [8].

$$MI = log \frac{P(w,r,w')}{P(w,*,*)P(*,r,w')}$$

Here, $P(w,r,w')$ is the probability of seeing *chat* aligned to *the* in a French-English parallel corpus, and $P(w,*,*)P(*,r,w')$ is the product of the probability of seeing *chat* aligned to any word in the corpus and the probability of seeing *the* aligned to any word in the corpus.

5 Evaluation

There are several evaluation methods available to assess lexico-semantic data. [7] distinguishes several. We decided to compare against a gold standard, because there is a large French synonym dictionary available. We evaluated our results on the *Dictionnaire Electronique des Synonymes* (DES, [26]), which is based on a compilation of seven French synonym dictionaries. It contains 49,149 nodes connected by 200,606 edges that connect synonymous words.

We compare our results to the syntax-based method for French by [3]. They present an explorative study of using distributional similarity to extract synonyms for French. They use two corpora: a 200 million-word corpus of newspaper text from *Le Monde*, and a 30 million-word corpus consisting of 515 twentieth century novels. Several syntactic relations are extracted.

The test set was chosen by looking at the pairs of candidate synonyms resulting from the syntax-based method that receive a score not lower than 0.16. This resulted in a list of approximately 1000 nouns. Of this list 950 can be found in the data of the alignment-based method. This list of 950 word constitutes the test set.

6 Results and Discussion

The results can be seen in Figure 3 and Figure 4. The x-axis indicates the threshold we set for the similarity score. For each candidate synonym the system calculates a similarity score. The dot at 0.20 in Figure 3 shows the average precision of all candidate synonyms for all testwords that have a similarity score of 0.20 or higher. Precision and recall are calculated as well as the coverage of the system at varying thresholds. Coverage indicates for how many of the testwords the system finds a synonym at the given similarity threshold.

Coverage of both systems decreases when the threshold for the similarity score is augmented. That is expected since not many words have candidate synonyms with a high similarity score. The alignment-based method never reaches 100% coverage. However, it should be noted that the test set was chosen in a way that favours the syntax-based method. The test set is composed of pairs of candidate synonyms resulting from the syntax-based method that are above the threshold 0.16. Thus, the coverage of the syntax-based method is 100% at 0.16. The coverage of the alignment-based method is approximately 70% for that threshold. However, the

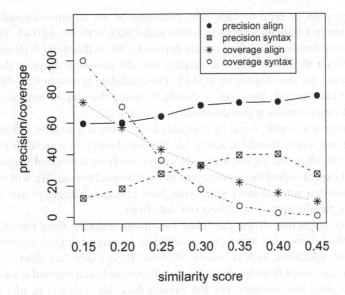

Fig. 3 Precision and coverage for the two methods at several thresholds of similarity score

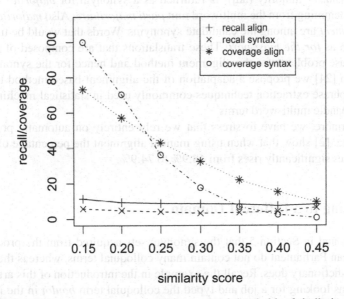

Fig. 4 Recall and coverage for the two methods at several thresholds of similarity score

coverage of the syntax-based method decreases more rapidly as the thresholds are raised. At threshold 0.45 the coverage of the syntax-based method is close to zero.

If we compare the precision of the candidate synonyms for both systems at the same level of coverage (50%) we see that the syntax-based method has a precision score of 25%, whereas the alignment-based method produces candidate synonyms with a precision of 60% to 65%. The precision of the alignment-based method ranges between a little under 60% to a little under 80% at threshold 0.45. The precision of the syntax-based method ranges between 10% at threshold 0.16 and a little under 40% for threshold 0.4. It is striking that the precision drops at the end of the line, when the threshold is set to 0.45. The candidate synonym with the highest scores are not the best. However, it should be noted that due to limited coverage (close to 0) the numbers at this threshold are unreliable.

With respect to recall, it can be concluded that there is a smaller difference between the two methods and the scores are less satisfactory in general. It should be noted that the dictionaries often include synonyms from colloquial language use. We do not expect to find these synonyms in the Europarl corpus. We will see in the next section that we find more synonyms from colloquial language use when we change the parallel corpus we collect our data from.

A closer inspection of the candidate synonyms resulting from the alignment-based method shows that many of the candidate synonyms judged incorrect are in fact valuable additions, such as *sinistre* 'disaster' for *accident* 'accident'.

Many errors stem from the fact that the alignment-based method does not take multiword units into account. For the French data this typically results in many related adjectives and adverbs being selected as candidate synonyms. For example, *majoritaire* 'majority (adj)' is returned as a synonym for *majorité* 'majority (noun)', stemming from the multiword unit *parti majoritaire*. Also *majoritairement* and *largement* are among the candidate synonyms. Words that would be translated in English as *for the most part*. These translations that are composed of multiple words cause problems for the alignment method and hence for the synonyms extracted. In [24] we propose a adaptation of the alignment-based method that uses standard phrase extraction techniques commonly used in statistical machine translation to handle multi-word terms.

Furthermore, we have to stress that we rely entirely on automatic processing of our data. [2] show that when using manual alignment the percentage of correct paraphrases significantly rises from 48.9% to 74.9%.

7   Porting to a Different Domain

We mentioned in Section 5 that the synonym sets acquired from the proceedings of European Parliament do not contain many colloquial terms, whereas the French synonym dictionary does. Recall the example in the introduction of this article, the student was looking for a job and typed the colloquial term *boulot* in the interface of the search engine.

To be able to find synonyms typical for every day language use, we need to give the system access to corpora of every day language use. The conversations in a large variety of movies are probably closer to the every day language use of an average French student than the proceeding of the European parliament. Instead of extracting translation pairs from the parallel corpus Europarl, we used a multilingual parallel corpus of movie subtitles, the OpenSubs corpus[31].

This method allows us to acquire synonyms that are specific to any particular domain, for example, in [24] we extended the alignment-based method to find medical term variants.

7.1 Materials and Methods

The OpenSubs corpus contains about 21 million aligned sentence fragments in 29 languages. We used all language pairs that include French, 23 language pairs in total. Still the corpus is much smaller than the Europarl corpus. The number of translation pairs (hapaxes excluded) we extract from this corpus is more than 50 times smaller than the number of translation pairs extracted from Europarl.

The domain is different from the domain of the Europarl corpus. There is a world of difference between the working day of a member of the European Parliament and the adventures of Nemo. Moreover, movie subtitles consist mainly of transcribed speech. In principle this is the same for the Europarl corpus. However these proceedings are edited and far less spontaneous than the speech data from the movies.

Another difference is the amount of pre-processing we applied. For the Europarl corpus we had access to lemma information for the words because, as we explained, the corpus was lemmatised. For the OpenSubs corpus we did not have access to lemma information and thus used the words instead. These experiments are therefore a good testbed for the usefulness of the method when there is limited data available and limited resources, as is the case for many resource-poor languages. We again applied a cell and row frequency cutoff of four and ten.

7.2 Evaluation of the Synonyms from the OpenSubs Corpus

Because the OpenSubs corpus has not been lemmatised, the resulting synonyms have all kinds of inflections. The synonym dictionary we use for the evaluation contains lemmas. This means that if we evaluate the synonyms on the dictionary it would not recognise the inflected wordforms (plurals and feminine forms) and count them as incorrect. We therefore converted every plural form to its singular form, using the Morphalou database from the : *Centre National de Ressources Textuelles et Lexicales*[2] and we converted feminine forms to their masculine counterpart. We then removed the doubles from the lists of candidate synonyms and ran the evaluations using the synonym dictionary.

Note that despite this pre-processing, the acquisition of synonyms from the non-lemmatised corpus of subtitles is a more difficult task than the acquisition of synonyms from the lemmatised Europarl corpus. Lemmatisation does not only make evaluation on the dictionary easier, it also reduces data sparseness. For a small corpus such as the OpenSubs corpus data sparseness is more problematic than for larger corpora. On the other hand this setting illustrates the performance of the alignment-based method in a setting that might be realistic for many languages: only a small amount of data and no lemmatiser is available. In the end, our aim is to use as little

---

[2] The resource is available from http://www.cnrtl.fr/lexiques/morphalou/

language-specific pre-processing as possible to assure a wide applicability of the method.

7.3 Results for the OpenSubs Corpus

The main reason for applying the method to a corpus from another domain is to find synonyms that are typical for that domain. From the examples given in Table 3 we get the impression that the synonyms stemming from the OpenSubs corpus contain more slang and colloquial language use, such as *nana* 'babe' as a candidate synonym for *fille* 'girl'. At the same time we see that the proceedings of the European parliament constitute a specific domain as well. In the context of the European parliament a friend is like a comrade and an ally, synonyms very specific to the the the particular domain, whereas in movies friends are pals and buddies. Similarly, the adjective *malade* for which the default translation would be 'ill', gets synonyms from the OpenSubs corpus like *fou* 'mad' and *dingue* 'crazy'

Table 3 Examples of candidate synonyms at the top-3 ranks for two corpora

Testword	Corpus			
ami	Europarl	amitié	camarade	allié
'friend'		'friendship'	'comrade'	'ally'
	Subtitles	copain	pote	amie
		'pal'	'buddy'	'girlfriend'
fille	Europarl	fillette	enfant	filial
'girl'		'small girl'	'child'	'relative to a daughter'
	Subtitles	fillette	nana	fiiie
		'small girl'	'babe'	(mistake in optical character recognition)
malade	Europarl	patient	maladie	souffrant
'ill'		'patient'	'illness'	'suffering'
	Subtitles	souffrant	fou	dingue
		'suffering'	'mad'	'crazy'

We also calculated precision, recall, and coverage for the synonyms acquired from the OpenSubs corpus as can be seen in Figure 5. The scores are not as good as when using the Europarl corpus. We expect that using the subtitle corpus leads to poorer performance. There are at least three reasons for this. The first reason is that the corpus is smaller. The second reason is that the subtitle corpus is more noisy than the Europarl corpus, for example due to mistakes in optical character recognition, as we see in Table 3. The last is the absence of lemmatisation. Still, despite the small amount of data, the noise and the absence of language-specific pre-processing the system is able to find synonyms for more than 20% of the words in the testset with a precision of around 40% at the lowest threshold, which is promising. It means that the method can be applied to any language and any domain for which there exists a relatively small, possibly noisy, parallel corpus, without the need of language-specific pre-processing. This is good news from the point of view of sense induction

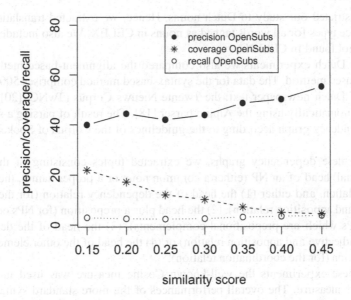

Fig. 5 Precision, recall and coverage for the synonyms stemming from the OpenSubs corpus

as well. The several senses a word has are often limited by a specific domain. We can deduce from the synonyms the word *malade* gets in Table 3 that in the context of the proceedings of the European Parliament the word gets the sense ill, whereas in the context of movie subtitles another sense of the word *malade*, namely the sense crazy, is apparent as well.

8 Porting to a Different Language

We explained in the introduction that the method easily ports to different languages. As an example we include results of applying the alignment-based method to Dutch taken from [21].

8.1 Materials and Methods

Instead of collecting translations for all French words, we collected translations for all Dutch words in the same parallel corpus used to acquire the French synonyms, the Europarl corpus. We post-processed the alignment results in various ways for Dutch. We applied a simple lemmatiser to the Dutch part of the bilingual translational co-occurrence types. For this we used two resources: CELEX, a linguistically annotated dictionary of English, Dutch, and German [1], and the Dutch snowball stemmer implementing a suffix-stripping algorithm based on the Porter stemmer. We removed word type links that include non-alphabetic characters and we transformed all characters to lower case.

We restricted our study to Dutch nouns. Hence, we extracted translational co-occurrence types for all words tagged as nouns in CELEX. We also included words that are not found in CELEX[3].

In the Dutch experiments we also compared the alignment-based method to a syntax-based method. The data for the syntax-based method comprises 500 million words of Dutch newspaper text: the Twente Nieuws Corpus (TwNC, [20]) that is parsed automatically using the Alpino parser [18]. The result of parsing a sentence is a dependency graph according to the guidelines of the Corpus of Spoken Dutch [17].

From these dependency graphs, we extracted tuples consisting of the (non-pronominal) head of an NP (either a common noun or a proper name), the dependency relation, and either (1) the head of the dependency relation (for the object, subject, and apposition relation), (2) the head plus a preposition (for NPs occurring inside PPs which are prepositional complements), (3) the head of the dependent (for the adjective and apposition relation) or (4) the head of the other elements of a coordination (for the coordination relation).

For these experiments the well-known Cosine measure was used instead of the Dice† measure. The overall performances of the more standard syntax-based method were highest when using Cosine, so comparative evaluations in [21] were done using Cosine. For the same reason cell and row cutoffs were set to two.

Evaluations for Dutch were done on a large test set of 3000 nouns selected from Dutch EuroWordNet. The test set is split up in equal amounts (1000) of high-frequency (HF), middle-frequency (MF) and low-frequency (LF) words. This was done to be able to study the effect of frequency on the performance of the system. We used the synsets in Dutch EuroWordnet [33] for the evaluation of the proposed synonyms.

8.2 Results on Synonym Extraction for Dutch

In Table 4 we see the comparative results on Dutch. The percentage of synonyms for the top-k candidate synonyms is given for the two methods and the three testsets. We see that 31.71% of the candidate synonyms produced by the alignment-based method as the most probable candidate synonym for the words in the high-frequency testset are indeed synonyms, whereas the syntax-based method only finds 21.31%.

Table 4 Percentage of synonyms over the k candidates for the alignment-based and syntax-based method for the three frequency bands

Method	HF		MF		LF	
	$k=1$	$k=5$	$k=1$	$k=5$	$k=1$	$k=5$
Alignment-based	31.71	19.16	29.26	16.20	28.00	16.22
Syntax-based	21.31	10.55	22.97	10.11	19.21	11.63

---

[3] Discarding these words would result in losing too much information. We assumed that many of them will be productive noun constructions.

The performance of the syntax-based method decreases rapidly when we go down the list of candidate synonyms, i.e. at higher values of k. For the high-frequency test set this is most apparent: From $k=1$ to $k=5$ the syntax-based method precision score is halved. At $k=5$ the alignment-based method receives still 2/3rd of the score at $k=1$. The syntax-based method retrieves about 2/3rd of the synonyms the alignment-based method retrieves for the high-frequency test set.

Although the results are less clear-cut than the results for french, they show a similar pattern. In spite of data sparseness, it is clear from Table 4 that the alignment-based method is better at finding synonyms than the syntax-based method.

Differences in performance between the experiments on Dutch and on French can be partly explained by the difference in the goldstandards used in the evaluation. The French dictionary contains 49,149 nodes connected by 200,606 synonym edges. EWN contains a total of 56,283 entries. However, the degree of synonymy is higher. For EWN the degree of synonymy is expressed by the ratio of senses per synset: 1.59. The ratio of edges per entry in the case of the DES is 4. A high degree of synonymy favours high precision scores.

In an evaluation with human judgements [23] showed that in 37% of the cases the majority of the subjects judged the synonyms proposed by the system to be correct even though they were not found to be synonyms in Dutch EuroWordnet. The results from human judgements lead us to believe that the method performs better than the scores in Table 4 indicate. Over and above, this indicates that we are able to extract automatically synonyms that are not yet covered by available resources.

Of course the differences in nature of the two languages also play their roles. Dutch uses single-word compounding, contrary to the majority of languages that use multiple words to describe a concept. For example, in English, compounds are mostly composed of two words orthographically, e.g *table cloth* and *hard disk* versus *database*. The Dutch word *slagroom* 'whipped cream' is a single word. These single-word compounds introduce errors in the word alignments, where *slagroom* is attached to either *whipped* or *cream*. French behaves like English in this respect.

9 Conclusions

We have shown that the alignment-based method outperforms the traditional syntax-based method for the task of automatic synonym acquisition by a very large margin on the task of French synonym acquisition. The precision is more than twice as high for the alignment-based method and it manages to find valuable additions not present in the large synonym dictionary on which it was evaluated. In addition, we showed that the method can be easily ported across languages and domains. We showed that we can retrieve synonyms of a different nature by using a corpus of movie subtitles instead of a corpus that consists of proceedings from the European parliament. The method can be easily adapted to a specific domain. Moreover, the method works reasonably well, even when using small, noisy corpora, and no language-specific pre-processing. This opens the way to the acquisition of synonyms for specialised domains and resource-poor languages. The method can be easily applied to other languages and results in similar performances. It compares favourably to the syntax-based methods for the acquisition of Dutch synonyms as is the case for French synonym acquisition.

Acknowledgements

Part of this work has received funding from the EU FP7 programme (FP7/2007-2013) under grant agreement nr 216594 (CLASSIC project: www.classic-project.org). It is based on research carried out in the project *Question Answering using Dependency Relations*, which is part of the research program for *Interactive Multimedia Information eXtraction*, IMIX, financed by NWO, the Dutch Organisation for Scientific Research.

References

1. Baayen, R., Piepenbrock, R., van Rijn, H.: The CELEX lexical database (CD-ROM). Linguistic Data Consortium, University of Pennsylvania, Philadelphia (1993)
2. Bannard, C., Callison-Burch, C.: Paraphrasing with bilingual parallel corpora. In: Proceedings of the annual Meeting of the Association for Computational Linguistics, ACL (2005)
3. Bourigault, D., Galy, E.: Analyse distributionnelle de corpus de langue générale et synonymie. In: Lorient, Actes des Journées de la Linguistique de Corpus, JLC (2005)
4. Brown, P.F., Della Pietra, S.A., Della Pietra, V.J., Mercer, R.L.: The mathematics of statistical machine translation: Parameter estimation. Computational Linguistics 19(2), 263–296 (1993)
5. Callison-Burch, C.: Syntactic constraints on paraphrases extracted from parallel corpora. In: Proceedings of EMNLP (2008)
6. Church, K.W., Hanks, P.: Word association norms, mutual information and lexicography. In: Proceedings of the Annual Conference of the Association of Computational Linguistics, ACL (1989)
7. Curran, J.: From distributional to semantic similarity. Ph.D. thesis, University of Edinburgh (2003)
8. Curran, J.R., Moens, M.: Improvements in automatic thesaurus extraction. In: Proceedings of the Conference on Empirical Methods in Natural Language Processing, EMNLP, pp. 222–229 (2002)
9. Dagan, I., Itai, A., Schwall, U.: Two languages are more informative than one. In: Proceedings of the Annual Meeting of the Association for Computational Linguistics, ACL (1991)
10. Dyvik, H.: Translations as semantic mirrors. In: Proceedings of Workshop Multilinguality in the Lexicon II (ECAI) (1998)
11. Harris, Z.S.: Mathematical structures of language. Wiley, Chichester (1968)
12. Hindle, D.: Noun classification from predicate-argument structures. In: Proceedings of the Annual Meeting of the Association of Computational Linguistics, ACL (1990)
13. Ide, N., Erjavec, T., Tufis, D.: Sense discrimination with parallel corpora. In: Proceedings of the ACL Workshop on Sense Disambiguation: Recent Successes and Future Directions (2002)

14. Koehn, P.: Europarl: A multilingual corpus for evaluation of machine translation (2003)
15. Lin, D.: Automatic retrieval and clustering of similar words. In: Proceedings of COL-ING/ACL (1998)
16. Lin, D., Zhao, S., Qin, L., Zhou, M.: Identifying synonyms among distributionally similar words. In: Proceedings of the International Joint Conference on Artificial Intelligence (IJCAI) (2003)
17. Moortgat, M., Schuurman, I., van der Wouden, T.: CGN syntactische annotatie, Internal Project Report Corpus Gesproken Nederlands (2000),
    http://lands.let.kun.nl/cgn
18. van Noord, G.: At last parsing is now operational. In: Actes de la 13eme Conference sur le Traitement Automatique des Langues Naturelles (2006)
19. Och, F.: GIZA++: Training of statistical translation models (2003),
    http://www.isi.edu/~och/GIZA++.html
20. Ordelman, R.: Twente nieuws corpus (TwNC). Parlevink Language Techonology Group. University of Twente (2002)
21. van der Plas, L.: Automatic lexico-semantic acquisition for question answering. Groningen dissertations in linguistics (2008)
22. van der Plas, L.: Automatic lexico-semantic acquisition for question answering. Ph.D. thesis, University of Groningen (2008)
23. van der Plas, L., Tiedemann, J.: Finding synonyms using automatic word alignment and measures of distributional similarity. In: Proceedings of COLING/ACL (2006)
24. van der PLas, L., Tiedemann, J.: Finding medical term variations using parallel corpora and distributional similarity. In: Proceedings of the Coling Workshop on Ontologies and Lexical Resources (2010)
25. van der Plas, L., Tiedemann, J., Manguin, J.L.: Extraction de synonymes à partir d'un corpus multilingue aligné. Actes des 5èmes Journées de Linguistique de Corpus à Lorient (2008)
26. Ploux, S., Manguin, J.: Dictionnaire électronique des synonymes français (1998, released 2007)
27. Resnik, P.: Selection and information, Unpublished doctoral thesis, University of Pennsylvania (1993)
28. Resnik, P., Yarowsky, D.: A perspective on word sense disambiguation methods and their evaluation. In: Proceedings of ACL SIGLEX Workshop on Tagging Text with Lexical Semantics: Why, What, and How? (1997)
29. Schmid, H.: Probabilistic part-of-speech tagging using decision trees. In: Proceedings of International Conference on New Methods in Language Processing, Manchester, UK, pp. 44–49 (1994), http://www.ims.uni-stuttgart.de/~schmid/
30. Shimota, M., Sumita, E.: Automatic paraphrasing based on parallel corpus for normalization. In: Proceedings of the International Conference on Language Resources and Evaluation, LREC (2002)
31. Tiedemann, J.: News from OPUS - A collection of multilingual parallel corpora with tools and interfaces. In: Nicolov, N., Bontcheva, K., Angelova, G., Mitkov, R. (eds.) Recent Advances in Natural Language Processing, vol. V, pp. 237–248. John Benjamins, Amsterdam (2009)
32. Tiedemann, J., Nygaard, L.: The OPUS corpus - parallel & free. In: Proceedings of the International Conference on Language Resources and Evaluation, LREC (2004)
33. Vossen, P.: EuroWordNet A Multilingual Database with Lexical Semantic Networks. Kluwer Academic Publishers, Dordrecht (1998)
34. Wu, H., Zhou, M.: Optimizing synonym extraction using monolingual and bilingual resources. In: Proceedings of the International Workshop on Paraphrasing: Paraphrase Acquisition and Applications, IWP (2003)

Linguistically-Based Reranking of Google's Snippets with GreG

Rodolfo Delmonte and Rocco Tripodi

Abstract. We present an experiment evaluating the contribution of a system called GReG for reranking the snippets returned by Google's search engine in the 10 hits presented to the user and captured by the use of Google's API. The evaluation aims at establishing whether or not the introduction of deep linguistic information may improve the accuracy of Google or rather it is the opposite case as maintained by the majority of people working in Information Retrieval and using a Bag Of Words approach. We used 900 questions and answers taken from TREC 8 and 9 competitions and execute three different types of evaluation: one without any linguistic aid; a second one with tagging and syntactic constituency contribution; another run with what we call Partial Logical Form. Even though GReG is still work in progress, it is possible to draw clear cut conclusions: adding linguistic information to the evaluation process of the best snippet that can answer a question improves enormously the performance. In another experiment we used the actual texts associated to the Q/A pairs distributed by one of TREC's participant and got even higher accuracy.

1 Introduction

We present an experiment run using Google API and a fully scaled version of GETARUNS, a system for text understanding (see Delmonte 2007; 2005), together with a modified algorithm for semantic evaluation presented in RTE3 (Recognizing Text Entailment) under the acronym of VENSES – Venice Semantic Evaluation System (Delmonte 2007). The aim of the experiment and of the new system that we called GReG (GETARUNS ReRANKS Google), is that of producing a reranking of the 10 hits presented by Google in the first page of a web search. Reranking is produced solely on the basis of the snippets associated to each link.

GReG uses a very "shallow" linguistic analysis by means of FSA (Finite State Automata) which nonetheless ends up with a fully instantiated sentence level

Rodolfo Delmonte · Rocco Tripodi
Department of Language Science
Università "Ca Foscari"
30123 – Venezia, Italy

V. Pallotta, A. Soro, and E. Vargiu (Eds.): Advances in DART, SCI 361, pp. 59–79.
springerlink.com   © Springer-Verlag Berlin Heidelberg 2011

syntactic constituency representation. In other words, we don't limit our analysis to the level of chunks but produce clause level and then sentence level structure, where grammatical functions have been marked on a totally bottom-up analysis and the subcategorization information associated to each governing predicate – verb, noun, adjective. More on this process in the sections below.

At the end of the parsing process, GReG produces a translation into a flat minimally recursive Partial Logical Form (hence PLF) where besides governing predicates – which are translated into corresponding lemmata – we use the actual words of the input text for all linguistic relations encoded in the syntactic structure. Eventually all matching processes are carried out coupling semantic similarity measures over the words involved, dependency labels and logical relations.

There is now general consensus on the usefulness of linguistic processing for Q/A open/closed domain tasks. However, the need to keep the processing to a feasible amount of CPU time has led many researchers into the idea that dependency parsing is the only technology able to cope with the task. In fact, dependency word level parsing can become a too poor linguistic representation in many relevant cases. Some of the problems have been overcome by introducing "equivalence paraphrases" which are used to account for syntactic variations (Wang et al. 2007; Bouma et al. 2005).

Some other problems are more related to semantic completeness and factitivity. We are referring here to the problem of recovery of implicit arguments, either by means of binding of syntactic variables or by attaching the appropriate label to underlying object of passive sentences, or again binding the subject of untensed clauses, like gerundives, participials or infinitives. The other problem is related to the need to account for the presence of modality and negation operators which may affect the truth of the answer recovered and thus jeopardize the correctness of the results. These problems are coped with at the level of text entailment evaluation.

In our system, we address different levels of representations – syntactic and (quasi) logical/semantic, and measure their contribution if any in comparison to a baseline keyword or bag of words computation. Together with linguistic representation, we also use semantic similarity evaluation techniques already introduced in RTE (Recognizing Textual Entailment) challenges which seem particularly adequate to measure the degree of semantic similarity and also semantic consistency or non-contradictoriness of the two linguistic descriptions to compare. This is partially also proposed by others (Wang et al. 2007) when they introduce the use of WordNet to do answer expansion.

Differently from the majority of the systems in this field, we don't use any training, nor do we adopt a specific statistical model. The reason being simply the fact that we want our system to be highly performing in every domain and this may be only guaranteed by a solid and robust linguistic architecture.

2 The VENSES System

VENSES is a reduced version of GETARUNS (Delmonte, 2007 and 2009), a complete system for text understanding developed at the Laboratory of Computational Linguistics of the University of Venice. The backbone of VENSES is LFG theory in its original version (Bresnan, 1982 and 2000). The system produces different levels of analysis, from syntax to discourse. However, three of them contribute most to the classification task: the lexico-semantic, the anaphora resolution and the deep semantic module. The architecture of the parser is commented in this section. It is a quite common pipeline: all the code runs in Prolog and is made up of manually built symbolic rules.

The system produces a c-structure representation by means of a cascade of augmented FSA (Finite State Automata), then it uses this output to map lexical information from a number of different lexica which however contain similar information related to verb/adjective and noun subcategorization. The mapping is done by splitting sentences into clauses which are main and subordinate clauses. Other clauses are computed in their embedded position and can be either complement or relative clauses. The output of the system is what we call AHDS (Augmented Head Dependent Structure) which is a fully indexed logical form, with Grammatical Relations and Semantic Roles. The inventory of semantic roles we use is however very small – 35, even though it is partly overlapping the one proposed in the first FrameNet project. We prefer to use generic roles rather than specific Frame Elements (FEs) because sense disambiguation at this stage of computation may not be effective. The reasons are many but we just indicate one: semantic processes can be computed thoroughly (i.e. anaphora resolution) without the use of highly specialized semantic classification and taking advantage only of tagging disambiguation and of generic semantic roles, as for instance has been done in VerbNet, Namebank and PropBank subcategorization project. In addition, position of constituents is usually sufficient to determine their Grammatical Function or Dependency relation. Anaphora resolution uses all these labels to separate human from non human coreference, as the pronominal system does not require any deeper specification.

2.1 The Parser

The architecture of the parser is commented in this section. It is a quite common pipeline: all the code runs in Prolog and is made up of manually built symbolic rules. Here below in Fig. 1 we show the modules of the shallow parser and its interconnections.

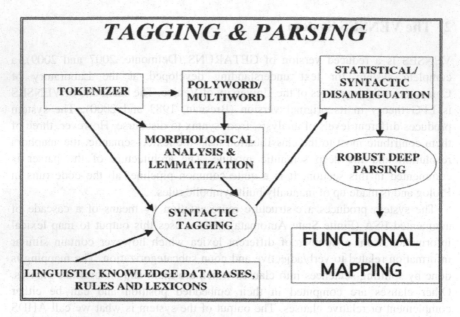

TAGGING & PARSING

Fig. 1 The shallow modules of VENSES

We defined our parser "mildly bottom-up" because the structure building process cycles on a procedure that collects constituents. This is done in three stages: at first chunks are built around semantic heads – verb, noun, adjective, adverbials. Then prepositions and verb particles are lumped together. In this phase, also adjectives are joined to the nominal head they modify. In a third phase, sentential structure information is added at all levels – main, relative clauses, complement clauses. In presence of conjunctions, different strategies are applied according to whether they are coordinating or subordinating conjunctions.

An important linguistic step is carried out during this pass: subcategorization information is used to tell complements – which will become arguments in the PLF – and adjuncts apart. Some piece of information is also offered by linear order: SUBJect NPs will usually occur before the verb and OBJect NP after. Constituent labels are then substituted by Grammatical Function labels. The recursive procedure has access to calls collecting constituents that identify preverbal Arguments and Adjuncts including the Subject if any: when the finite verb is found the parser is hampered from accessing the same preverbal portion of the algorithm and switches to the second half of it where Object NPs, Clauses and other complements and adjuncts may be parsed. Punctuation marks are also collected during the process and are used to organize the list of arguments and adjuncts into tentative clauses. This is shown in Fig. 2 below.

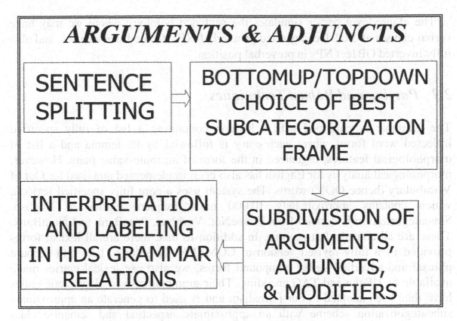

Fig. 2 Assignment of Grammatical Relations and Semantic Roles

The clause builder looks for two elements in the input list: the presence of the verb-complex and punctuation marks, starting from the idea that clauses must contain a finite verb complex: dangling constituents will be adjoined to their left adjacent clause, by the clause interpreter after failure while trying to interpret each clause separately. The clause-level interpretation procedure interprets clauses on the basis of lexical properties of the governing verb. This is often non available in snippets. So in many cases, sentence fragments are built.

If the parser does not detect any of the previous structures, control is passed to the bottom-up/top-down parser, where the recursive call simulates the subdivision of structural levels in a grammar: all sentential fronted constituents are taken at the CP level and the IP (now TP) level is where the SUBJect NP must be computed or else the SUBJect NP may be in postverbal position with Locative Inversion structures, or again it might be a subjectless coordinate clause. Then again a number of ADJuncts may be present between SUBJect and verb, such as adverbials and parentheticals. When this level is left, the parser is expecting a verb in the input string. This can be a finite verb complex with a number of internal constituents, but the first item must be definitely a verb. After the (complex) verb has been successfully built, the parser looks for complements: the search is restricted by lexical information. If a copulative verb has been taken, the constituent built will be labelled accordingly as XCOMP where X may be one of the lexical heads, P,N,A,Adv.

The clause-level parser simulates the sentence typology where we may have verbal clauses as SUBJect, Inverted postverbal NPs, fronted that-clauses, and also fully inverted OBJect NPs in preverbal position.

2.2 Parsing and Robust Techniques

The grammar is equipped with a lexicon containing a list of fully specified inflected word forms where each entry is followed by its lemma and a list of morphological features, organized in the form of attribute-value pairs. However, morphological analysis for English has also been implemented and used for Out of Vocabulary (hence OOV) words. The system uses a core fully specified lexicon, which contains approximately 10,000 most frequent entries of English. Subcategorization is derived from FrameNet, VerbNet, PropBank and NomBank. These are all consulted at runtime. In addition to that, there are all lexical forms provided by a fully revised version of COMLEX. In order to take into account phrasal and adverbial verbal compound forms, we also use lexical entries made available by UPenn and TAG encoding. Their grammatical verbal syntactic codes have then been adapted to our formalism and is used to generate an approximate subcategorization scheme with an approximate aspectual and semantic class associated to it – some information is derived from LCS (Lexical Conceptual Structure) from the University of Maryland (see Traum et al.). Semantic inherent features for OOV words, be they nouns, verbs, adjectives or adverbs, are provided by a fully revised version of WordNet – 270,000 lexical entries - in which we used 75 semantic classes. In addition to that we have a number of gazetteers and proper nouns lists including Arabic names, which amount to an additional 400,000 fully classified lexical entries.

2.3 The Anaphora Resolution Module

The AHDS structure is passed to and used by a full-fledged module for pronominal and anaphora resolution, which is in turn split into *two submodules*.

The resolution procedure takes care only of third person pronouns of all kinds – reciprocals, reflexives, possessive and personal. Its mechanisms are quite complex, as described in (Delmonte et al., 2006). The *first submodule* basically treats all pronouns at sentence level – that is, taking into account their position – and if they are left free, they receive the annotation "external". If they are bound, they are associated to an antecedent's index; else they might also be interpreted as expletives, i.e. they receive a label that prevents the following submodule to consider them for further computation. The *second submodule* receives as input the external pronouns, and tries to find an antecedent in the previous stretch of text or discourse. To do that, the systems computes a *topic hierarchy* that is built following suggestions by (Sidner and Grosz, 1986) and is used in a centering-like manner.

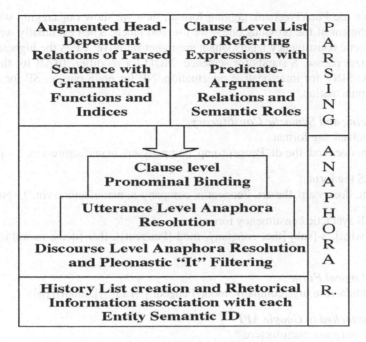

Augmented Head-Dependent Relations of Parsed Sentence with Grammatical Functions and Indices	Clause Level List of Referring Expressions with Predicate-Argument Relations and Semantic Roles	P A R S I N G
Clause level Pronominal Binding		A N A P H O R A R.
Utterance Level Anaphora Resolution		
Discourse Level Anaphora Resolution and Pleonastic "It" Filtering		
History List creation and Rhetorical Information association with each Entity Semantic ID		

Fig. 3 Information and modules of the Anaphora Resolution Algorithm

2.4 A Walkthrough Example

We now present three examples taken from TREC8 question/answer set, no. 3, 193, 195, corresponding respectively to ours 1,2,3. For each question we add the answer and then we show the output of tagging in PennTreebank format, then follows our enriched tagset and then the syntactic constituency structure produced by the parser and the grammatical labels. Eventually, we show the Partial Logical Form where the question word has been omitted. The question word will be transformed into its corresponding semantic type. In some cases, it can be reinserted in the analysis when the matching takes place and may appear in the other level of representation we present which is constituted by the Query in answer form passed to Google. Question words are always computed as argument or adjunct of the main predicate, so GReG will add a further match with the input snippets constituted by the semantic types of the wh- words. One such type is visible in question no.3 when the concept "AUTHOR" is automatically added by GReG in front of the verb and after the star. More on answer typing below.

(1) What does Peugeot company manufacture? – Cars
(2) Who was the 16th President of the United States? – Lincoln
(3) Who wrote "Dubliners"? – James Joyce

Here below are the analyses where we highlight the various levels of linguistic representation relevant for our experiment only – except for the default word level. We use PennTreebank tagging format, then we show our tagging which is an enrichment of the previous one – we use 100 labels. Then eventually we show the syntactic constituency labels which are rather intuitive: CP is the highest complementizer phrase, S stands for sentence, MOD for modifier, IBAR for the verb complex, SINT for interrogative punctuation, NP for noun phrase, SP for parenthetical punctuation.

(1) Tagging and Syntactic Constituency
PennTreebank tag format
what-wp, does-md, the-dt, Peugeot-nnp, company-nn, manufacture-vin, ? – pun

VENSES tag format
[what-int, does-vsup, the-art, Peugeot-n, company-n, manufacture-vin, ? - puntint]

VENSES syntactic constituency format
cp-[cp-[what], s-[subj-[the, company, mod-[Peugeot]], ibar-[does, manufacture]], sint-[?]]

Partial Logical Form
pred(manufacture)arg([company, mod([Peugeot])]) adj([[], mod([[]])])

Query launched to Google API
Peugeot company manufacture *

(2) Tagging and Syntactic Constituency
PennTreebank tag format
who-wp,   was-vbd,   the-dt,   16th-cd,   President-nnp,   of-in,   the-dt, United_States-nnp, ? – pun

VENSES tag format
[who-int,   was-vc,   the-art,   16th-num,   President-n,   of-p,   the-art, United_States-n, ? - puntint]

VENSES syntactic constituency format
sint-[cp-[who], ibar-[was], np-[the, 16th, President, mod-[of, the, United_States]], sint-[?]]

Partial Logical Form
[pred(be) arg([President, mod([united, States, 16th])]) adj([])]

Query launched to Google API
United States 16th President was *

(3) Tagging and Syntactic Constituency
PennTreebank tag format
who-wp, wrote-vbd_vbn, "-pun, Dubliners-nns, "-pun, ? - pun

VENSES tag format
[who-int, wrote-vt, "-par, Dubliners-n, "-par, ? - puntint]

VENSES syntactic constituency format
cp-[cp-[who], ibar-[wrote], sp-["], np-[Dubliners], sp-["], sint-[?]]

Partial Logical Form
pred(write) arg([Dubliners, mod([])]) adj([])

Query launched to Google API
* author wrote Dubliners

3 Semantics and Deep Processing

The output of the anaphora resolution module is used by the semantic module to substitute the pronoun's head with the antecedent's head. After this operation, the module produces Predicate-Argument Structures or PAS on the basis of previously produced Logical Forms. PAS are produced for each clause and they separate obligatory from non-obligatory arguments, and these from adjuncts and modifiers. Some adjuncts, like spatiotemporal locations, are only bound at propositional level. This module produces also a representation at propositional level, which for simplicity is just a simple vector of information containing 15 different slots, each one devoted to contain a different piece of semantic information. We encode the following items: modality, negation, focussing intensifiers/ diminishers, manner adjuncts, diathesis, auxiliaries, clause dependency if any from a higher governing predicate – this is the case for infinitivals and gerundives – and eventually a subordinator if any.

As said above, the idea is to try to verify whether deeper linguistic processing can contribute to question answering. As will be shown in the following tables, Google's search on the web has high recall in general: almost 90% of the answers are present in the first ten results presented to the user. However, we wanted to assume a much stricter scenario closer in a sense to TREC's tasks. To simulate a TREC task as close as possible we decided that only the first two snippets can be regarded a positive result for the user. Thus, everything that is contained in any of the following snippets will be computed as a negative result. We take two snippets to be similar to one sentence which is then the basis for the actual answer string.

The decision to regard the first two snippets as distinctive for the experiment is twofold. On the one side we would like to simulate as close as possible a TREC Q/A task, where however rather than presenting precise answers, the system is required to present the sentence/snippet containing it. The other reason is practical or empirical and is to keep the experiment user centered: user's attention should not be forced to spend energy in a tentative search for the right link. Focussing attention to only two snippets and two links will greatly facilitate the user. In this way, GReG could be regarded as an attempt at improving Google's search strategies and tools.

In order to evaluate the contribution of different levels of computation and thus get empirical evidence that a deep linguistically-based approach is worth while trying, we organized the experiment into a set of concentric layers of computation and evaluation as follows:

- at the bottom level of computation we situated what we call the "default semantic matching procedure". This procedure is used by all the remaining higher
level of computation and thus it is easy to separate its contribution from the overall evaluation;

- the default evaluation takes input from the first two processes, tokenization &
multiword creation plus sentence splitting. Again these procedures are quite standard and straightforward to compute. So we want to assume that the results are
easily reproducible as well as the experiment itself;

- the following higher level of computation may be regarded partly system dependent, but most of it is easily reproducible using off-the-shelf algorithms made
available for English by research centers all over the world. It regards tagging and
context-free PennTreebank-like phrase-structure syntactic representation as well
as dependency parsing. Here we consider not only words, but word-tag pairs and
word-as-head of constituent N pairs. We also take into account their grammatical
function label;

- the highest level is constituted by what we call Partial Logical Form, which
builds a structure containing a Predicate and a set of Arguments and Adjuncts
each headed by a different functor. In turn each such structure can contain Modifiers. Each PLF can contain other PLFs recursively embedded with the same structure. More on this below. This can also be reproduced by algorithms available
off-the-shelf at the DELPH-IN website.

3.1 Default Semantic Matching Procedure

This is what constitutes the closest process to the BOWs approach we can con-
ceive of. We compare every word contained in the Question with every word
contained in each snippet and we only compare content words. Stopwords are
deleted.

We match both simple words and multiwords. Multiwords are created on the
basis of lexical information already available for the majority of the cases. The
system however is allowed to guess the presence of a multiword from the information attached to the adjacent words and again made available in our dictionaries. If
the system recognizes the current word as a word starting with uppercase letter
and corresponding to one of the first names listed in one of our dictionary, it will
try to concatenate this word to the following and try at first a match. If the match
fails the concatenated word is accepted as a legitimate continuation – i.e. the name
– only in case it starts by uppercase letter. Similar checking procedures have been
set up for other NEs like universities, research centres, business related institutions
etc. In sum, the system tries to individuate all NEs on the basis of the information
stored and some heuristic inferential mechanism.

According to the type of NE we will licence a match of a simple word with a
multiword in different ways: person names need to match at least the final part
of the multiword, or the name institutions, locations etc. need to match as a
whole.

3.2   Tags and Syntactic Heads

The second level of evaluation takes as input the information made available by the tagger and the parser. We decided to use the same approach reported in the challenges called RTE where the systems participating could present more than one run and use different techniques of evaluation. The final task was – and is – that of evaluating the semantic similarity between the question and the input snippets made available by Google. However, there is a marked difference to be taken into account and is the fact that in RTE questions where turned into a fully semantically complete assertion; on the contrary, in our case we are left with a question word – applies to wh- questions - to be transformed into the most likely linguistic description that can be associated with the rest of the utterance. As most systems participating in TREC challenge have done, the question has to be rephrased in order to predict the possible structure and words contained in the answer, on the basis of the question word and overall input utterance. Some of the questions contained in the TREC list do not actually constitute wh- questions (factoid or list), but are rather imperatives or jussive utterances, which tell the system – and Google – to "describe" or to "name" some linguistic item specified in the following portion of the utterance.

As others have previously done, we classify all wh- words into semantic types and provide substitute words to be place in the appropriate sentence position in order to simulate as close as possible the answer. In other cases the semantic type will be used to trigger the appropriate general concept associated to the corresponding word matched in the snippet. In particular, whenever a number is required, be it a date, or other, the type QUANTITY is used to trigger the appropriate type. We distinguish between: DISTANCE, REPEAT, DURATION, DATE, POPULATION and a generic QUANTITY. In the latter case, however, a specific procedure checks for special cases of quantity definition, which may be SHARP, LESS, MORE, ABOUT, INCLUDES. Each subtype will compute the similarity accordingly.

However, this is only done in one of the modalities in which the experiment has been run. In the other modality, Google receives the actual words contained in the question.

As to experiment itself, and in particular to the matching procedure we set up, the wh- word is never used to match with the snippets. Rather we use the actual wh- words to evaluated negatively snippets containing them. In this way, we prevent similar and identical questions contained in a snippet and pointed by a link to receive a high score. We noticed that Google is unable to detect such mismatches.

We decided to use tag-word pairs in order to capture part of the contextual meaning associated to a given word. Also in the case of pairs word-as-head-of-constituent/ constituent label we wanted to capture part of the contextual import of a word in a structural representation and thus its syntactic and semantic relevance in the structure. As will be clear in the following section, this is different from what is being represented in a Logical Form for how partial it may be.

3.3 Partial Logical Form and Relations

The previous match intended to compare words as part of a structure of dependencies where heads played a more relevant role than non-heads, and thus were privileged. In the higher level match what we wanted to check was the possible relations intervening between words: in this case, matching regarded two words at a time in a hierarchy. The first and most relevant word was the PREDicate governing a given piece of PLF. The PRED can be the actual predicate governing at sentence level, with arguments and adjuncts, or it can be just the predicate of any of the Arguments/Adjuncts which in turn governs their modifiers.

Matching is at first applied to two predicates and if it succeeds, it is extended to the contents of the Argument or the Adjunct. In other words, if it is relations that this evaluation should measure, any such relations has to involve at least two linguistic elements of the PLF representation under analysis.

Another important matching procedure applied to the snippet is constituted by a check of the verbal complex – i.e. all linguistic elements constituting the verb structure: besides the inflected verb, the auxiliary, the modal, the negation, the possible adverbials and similar adjunct structures positioned in between the auxiliary/modal and the untensed main verb. We regard the verbal compound as the carrier of semantically important information to be validated at propositional level. However, seen the subdivision of tasks, we assume that we can be satisfied by applying a partial match. This verbal complex match is meant to ascertain whether the question and the answer contain positive polarity items; and in case they contain negative polarity items then they should be both containing one such item –not to convey contradictory information. It is also important to check whether the two verbal complexes are factitive or not: this is checked by detecting the presence of opaque or modality operators in the verb complex and at propositional adjunct level. This second possibility is matched carefully.

4 Greg's Algorithm in Some Detail

The reason for introducing and explaining the code that GREG uses is twofold: on the one side, we want to show how a system using linguistic rules is organized; on the other side, we believe it important to discuss in depth the various steps of the computations.

```
greg(NoFr, Input):-
        venses_parse(NoFr, Input,Alls),
        parseq(Input, Alls, Query),
        getsnippets(Input,Query).
```

The main call is used by GREG to parse the question and produces the prospective answer. Then it calls Google and passes the actual text of the question or its transformed version. Here it parses all the snippets and then evlauates the resulting LF structures matching each output with the correct answer.

```
venses_parse(NoFr, Sent,[Tags,Costs,AHDS]):-
        googmults(Sent,NewSent,Mults),
        googletagging_disamb(NoFr,NewSent,Tensed,Tags,Costs, AHDS).
parseq(Input, Alls, Query):-
        extract_query(Alls,Query,Qword,Verb,Passive),
        append_qwords(Costs,Query,Passive,ExAnswer,Verb,LF).
```

We list here below the procedures for extract_query and the those related to append_qwords.

```
extract_query(Alls,Query,Q1,Verb,Passive):-
        Alls=[Tags, Costs, AHDS],
        recoverqword(Cp,Output,RestCosts,Q1),
        extractquery(RestCosts,Query,Verb, Passive,Q1).
        extractquery(Costs,[Query,LF],Verb,Passive,Q1):-
        extrq1(Q1,Sent,Subj,Verb,Passive),
        extrq2(Q1,Sent,Obj),
        extrq3(Sent,Adjs),
        getpredverb(Verb,Pred),
        extrq4(Pred,Passive,Sent,Vcomps,Arg),
        buildanswer(Passive,Verb,Pred,Subj,Obj,NAdjs,Vcomps,
                Arg,Queries,LF).
```

```
buildanswer(Passive,Verb,Pred,Subj,Obj,Adjs,Vcomp,Arg,Answer1,LF):-
        buildansw(Subj,Verb,Obj,Answer),
        checkpassivequery(Passive,Answer,Adjs,Answer1),
        lfsN(Pred,Subj,Obj,Adjs,LF).
```

```
lfsN(Pred,Subj,Obj,Adjs,LF):-
        getargsmods(Subj,Arg1,Mod1),
        getargsmods(Obj,Arg2,Mod2),
        getargsmods(Adjs,Adj,Mod3),
        LF=[pred(Pred),arg([Arg1,mod(Mod1)],
                arg([Arg2,mod(Mod2)]),adj([])]),adj([Hea,mod(Adj)])].
```

The type of q-word may be responsible for the choice of the semantic equivalent to be used in the search for the appropriate answer, and this is something that must be carefully analysed. What is usually done, is associating a semantic tag that classifies the q-word and helps the search.

```
append_qwords(Tags,[Queries,LF],Passive,Qword,Verb,ExpAnswer,LF):-
        filterqueriesngs(Queries,NQuery),
        append_qword(Tags,NQuery,Passive,Qword,Verb,ExpAnswer,LF).
```

append_qword ---> performs different actions according to the Qword and the Verb of the current question. Also notice that in the case of WHO factoid questions, the system checks for the presence of PASSIVE mood. It also uses WordNet synsets and derivations to insert other very close synonyms and also use the corresponding AGENTive word, as for instance in the case of a verb like kill→ killer, for write we use AUTHOR, etc.

Qword=why --> append(Query,[because],ExpAnswer)

Qword=how --> searchsynsverb(Verb,I), append(Query,I, ExpAnswer)

Qword=how_far --> composequery(distance,Verb,Query,ExpAnswer)

Qword=how_often --> composequery(repeat,Verb, Query,ExpAnswer)

Qword=how_long --> composequery(duration,Verb, Query,ExpAnswer)

Qword=how_big --> composequery(quantity,Verb, Query,ExpAnswer)

Qword=how_many & Verb=live--> composequery(population,Verb, Query, ExpAnswer)

Qword=when --> searchsynsverb(Verb,I),

appendOr(Query,I,Quer), append(Quer,[in, *], ExpAnswer)

Qword=who --> Passive=passive, append(Query,[*],Quer),

searchsynsverb(Verb,I), append(Quer,I,ExpAnswer)

;

Passive=active, append(Query,[by, *],Quer),

searchsynsverb(Verb,I),append(Quer,I, ExpAnswer)

The content of the prospective answer is sent to Google and the snippets are collected and analysed with the call GETSNIPPETS.

getsnippets(Input,Query):-
```
        googleToList(Query,Texts),
        convert_answers(Texts,Codes),
        qopen_file(1, Codes, Stats),
        google_parse(NoFr, Text, Out, N, Input, Alls).
```

Eventually, here is the final part of the algorithm, that takes all analyses produced for each snippet and evaluates its relevance by comparing it with the prospective answer.

evaluation(Input,Query, Alls):-
```
        Alls=[Tags, Costs, AHDS,LF],
evaloutssynt(Qword-Pred-Query,Costs,Input,OutStr,Output,Ev1,Tags),
evaloutssems(Qword-Pred-Query,LF,OutStr,Output,Ev1,Ev2),
extract_eval(Text,Tags,Ev2,Evsort,Pol1,Pol2,Pol3).
```

stage 1, extract_eval checks to see whether the best snippets resulting from GReG's computation coincides with Google's (match_google) and the result is

reported in the index Pol1 which can assume three values: 1, complete conci-
dence; 2, partial coincidence (GReG's best choice is not ranked the same); 0, no
coincidence. Notice that if Evsort=[] then the value of index Pol1 is set to 0.

stage 2, extract_eval searches for the presence of the answer in the snippets
(match_answer) and writes the output in Bests which is a list of terms each one
contains a value of the search and the snippet index.

stage 3, extract_eval checks to see whether the snippets containing the answer coincide with the best choices resulting from GReG's computation (search_answer).
Again the result is reported in an index Pol3, which may assume three values as
before.

```
extractevalpols(Text,Tagsall,Ev2,Evsort,P1,P2,P3):-
        (match_google(First,Second,P1)
        ;
        match_google(First,[],P1)
        ;
        P1=0),
        match_answer(Text,Ev2,Tagsall,Bests,Pol2),
        search_answer(Bests,Evsort,Pol3).
```

match_google :- this predicate takes as input what GReG has computed as best
two snippets and verify whether they coincide with Google's choice.

```
match_answer(Text,Ev2,Tagsall,Bests,Pol2):-
        assessanswer(TextAns,Pol,NewAns),
        matchtagsans(NewAns,Pol,Tagsall,Evs),
        select_best(Po,NewAns,Evs,Bests).
```

match_answer :- this predicate takes the answer and does the following steps:

- assessanswer(TextAns,Pol,NewAns)

it assesses the answer in Pol where it encodes whether the content of the answer is
constituted by a numerical answer or not - this is done by the predicate
evalintg(TextAns,Pol), where Pol can assume six different values: no (the answer
is not numerical); sharp (the answer is numerical and is constituted by a precise
value); less (the numerical value is less than a certain amount); more (the numerical value is more than a certain amount); about (the numerical value is about a
certain amount); and finally includes (the numerical value is included in a range)
in NewAns is contained the extended version of some abbreviation present in the
answer.

matchtagsans :- on the basis of the NewAns and of Pol, it matches each word and
tag contained in the list Tagsall, where each snippet is compared to the answer and

produces a first evaluation which basically counts how many words are matched in Evs. The main matching predicate is constr_main_head_roles(T0,T1), which checks whether two words are

- identical,
- synonyms,
- belong to the same lexical semantic field as listed by Roget's Thesaurus
- are one the derivation of the other
- are morphologically similar.

Answer Matching Problems: problems may arise from essentially three types of data:

- numerical data can be reported in an abbreviated format or in an approximate quantity manner
- numerical data which are related to some measure or some currency indicator
- person names which need to be checked completely, i.e. name and surname

5   Evaluation and Discussion

The evaluation will focus on a subset of the questions used in TREC made up of 900 question/answers pairs made available by NIST and produces the following data:

- how many times the answer is contained in the 10 best candidates retrieved by Google;
- how many times the answer is ranked by Google in the first two links – actually we will be using only snippets (first two half links);
- as a side-effect, we also know how many times the answer is not contained in the 10 best candidates and is not ranked in the first two links;
- how many times GReG finds the answer and reranks it in the first two snippets;
- how much contribution is obtained by the use of syntactic information;
- how much contribution is obtained by means of LF, which works on top of syntactic representation;
- how much contribution is obtained by modeling the possible answer from the question, also introducing some meta operator – we use OR and the *.

The metric we adopt is very similar to the one proposed in Bouma et al. 2005, where they use what they call d-score for dependency relations evaluation, and t-score for syntactic dependency evaluation. The additional information we compute is related to the way we match head words or predicates, which are checked not only for equivalence but also for semantic similarity using the set of semantic relations made available by WordNet; the two words may also belong to the same semantic field as computed by Roget's Thesaurus and other similar lexical resources. Eventually, we compute accuracy measures by means of the usual Recall/Precision formula.

Here below we show the output of GReG in relation to one of the three questions presented above, question n.2

google7
Evaluation Score from Words and Tags : 31
Evaluation Score from Syntactic Constituent-Heads : 62
Evaluation Score from Partial Logical Form : 62
google8
Evaluation Score from Words and Tags : 35
Evaluation Score from Syntactic Constituent-Heads: 70
Evaluation Score from Partial Logical Form : 0
google9
Evaluation Score from Words and Tags : 33
Evaluation Score from Syntactic Constituent-Heads : 66
Evaluation Score from Partial Logical Form : 66

Snippet No. google9
16th President of the United States (March 4, 1861 to April 15, 1865). Nicknames: " Honest Abe " " Illinois Rail - Splitter ". Born : February 12, 1809 , . . .

Snippet No. google7
Abraham Lincoln , 16th President of the United States of America, 1864, Published 1901 Giclee Print by Francis Bicknell Carpenter - at AllPosters . com .

The right answer is : Lincoln

Google's best snippets containing the right answer are:

google8
Who was the 16th president of the united states ? pissaholic . . . . Abraham Lincoln was the Sixteenth President of the United States between 1861 - 1865 . . .

google7
Abraham Lincoln, 16th President of the United States of America, 1864, Published 1901 Giclee Print by Francis Bicknell Carpenter - at AllPosters . com .

Google's best answer partially coincides with GReG.

****

Passing Questions to Google filtered by GReG's analysis produced a positive result in that 755 questions contained the answer in the 10 best links. On the contrary, passing Questions to Google as is, produces as a result that only in 694 questions contain the answer in the 10 best links. In other words, GReG's analysis of the question triggers best results from Google, in fact improving the ability of Google to search for the answer and select it in the best 10 links.

In fact, Google exploits the linear order of words contained in the question. So in case there is some mismatch the answer is not readily found or perhaps is available further down in the list of links. In Table 1 we report in the first two rows Google's overall results divided up between the case in which the snippets contains or does not contain the right answer. In the second half of Table 1 we report data related only to the 755 positive results, where Google has found the right answer in its ten snippets. As can be noticed, only 28% of the positive cases are presented in first two snippets.

Table 1 Google outputs with and without the intervention of GReG's question analysis

	With GReG's preanalysis	Without GReG's anal.
Google's 10 Best links contain the answer	755 83.89%	694 77.12%
Google's 10 Best links do not contain the answer	145 16.11%	206 22.8%
Google Rank answer in first 2 snippets	216 28.61%	168 22.25%
Google Rank answer not in first 2 snippets	539 71.39%	587 77.75%

Table 2 GReG's outputs at different levels of linguistic complexity

GReG reranks the answer in first 2 snippets	Only word match	Tagging and Syntactic heads	Partial Logical Form
With GReG's analysis	375 49.67%	514 68.08%	543 71.92%
Without GReG's analysis	406 53.77%	493 65.30%	495 65.56%

5.1 Discussion

The conclusions we may safely draw is the clear improvements in performance of the system when deep linguistic information is introduced in the evaluation process. In particular, when comparing the contribution of PLF to the reranking process we see that there is a clear improvement: in the case of reranking without GReG's question analysis there is a slight but clear improvement in the final accuracy. Also, when GReG is used to preanalyse the question to pass to Google the contribution of PLF is always apparent. The overall data speak in favour of both preanalysing the question and using more linguistic processing.

If we consider Google's behaviour to the two inputs, the one with actual questions and the one with prospective answers we see that the best results are again obtained when the preanalysis is used; also the number of appropriate candidates – the recall - containing the answer increases remarkably when using GReG preprocessing (83% vs. 77%).

5.2 GReG and Question-Answering from Text

In order to verify the ability of our system to extract answers from real text we organized an experiment which used the same 900 question run this time against the texts made available by TREC participants. These texts have two indices at the beginning of each record line indicating respectively the question number which they should be able to answer, and the second an abbreviation containing the initial letters of the newspaper name and the date. In fact each record has been extracted by means of automatic splitting algorithms which have really messed up the whole text. In addition, the text itself has been manipulated to produce tokens which however do not in the least correspond to actual words of current orthographic forms in real newspapers. So it took us quite a lot of work to normalize the texts (5Mb.) to make them as close as possible to actual orthography.

Eventually, when we launched our system it was clear that the higher linguistic component could not possibly be used. The reason is quite simple: texts are intermingled with lists of items, names and also with tables. Since there is no principled way to tell these apart from actual texts with sentential structure, we decided to use only tagging and chunking.

We also had to change the experimental setup we used with Google snippets: in this case, since we had to manipulate quite complex structures and the choice was much more noisy, we raised our candidate set from two to four best candidates. In particular we did the following changes:

- we choose all the text stretches – usually corresponding to sentences - containing the answer/s and ranked them according to their semantic similarity;
- then, we compared and evaluated these best choices with the best candidates produced by our analyses;
- we evaluated to success every time one of our four best candidates was contained in the set of best choices containing the answer;
- otherwise we evaluated to failure.

In total, we ran 882 questions because some answers did not have the corresponding texts. Results obtained after a first and only run – which took 1 day to complete on an HP workstation with 5GB of RAM, 4 Dual Core Intel processors, under Linux Ubuntu – were quite high in comparison with the previous ones, and are reported here below:

Table 3 GReG's results with TREC8/9 texts

GReG finds the answer in first 4 text stretches	Tagging and Syntactic heads
Without GReG's analysis	684 / 882 77.55%

With respect to the favourable results, we need to consider that using texts provides a comparatively higher quantity of linguistic material to evaluate and so it favours better results.

6 Conclusions and Future Work

Overall, we believe to have shown the validity of our approach and the usefulness of deep linguistically-based evaluation methods when compared with shallower approaches. Structural and relational information constitutes a very powerful addition to simple tagging or just word level semantic similarity measures.

We intend to improve both the question translation into the appropriate format for Google, and the rules underlying the transduction of the Syntactic Structures into a Partial Logical Form. Then we will run the experiments again. Considering the limitations imposed by Google on the total number of questions to submit to the search engine per day, we are unable to increase the number of questions to be used in a single run.

We also intend to run GReG version for text Q/A this time with question rephrasing. We would also like to attempt using PLF with all the text stretches, after excluding manually all tables and lists. We are aware of the fact that this would constitute a somewhat contrived and unnatural way of coping of unrestricted text processing. At the same time we need to check whether the improvements we obtained with snippets are attested by the analysis of complete texts.

References

Bresnan, J.: Lexical-Functional Syntax. Blackwell, Malden (2000)
    ComLex, http://nlp.cs.nyu.edu/comlex
Baker, C.F., Fillmore, C.J., Lowe, J.B.: The Berkeley FrameNet project. In: Proceedings of
    COLING-ACL 1998, Montreal, Canada (1998)
Ellsworth, M., Erk, K., Kingsbury, P., Pado, S.: PropBank, SALSA, and FrameNet: How
    design determines product. In: Proceedings of the LREC 2004 Workshop on Building
    Lexical Resources from Semantically Annotated Corpora, Lisbon (2004)
Fellbaum, C. (ed.): WordNet: An Electronic Lexical Database. MIT Press, Cambridge
    (1998)
Delmonte, R. (ed.): Computational Linguistic Text Processing – Logical Form, Semantic
    Interpretation, Discourse Relations and Question Answering. Nova Science Publishers,
    New York (2007)
Delmonte, R.: Computational Linguistic Text Processing – Lexicon, Grammar, Parsing and
    Anaphora Resolution. Nova Science Publishers, New York (2009)
Delmonte, R.: Deep & Shallow Linguistically Based Parsing. In: Delmonte, R., Di Sciullo,
    A.M. (eds.) UG and External Systems, pp. 335–374. John Benjamins, Amsterdam
    (2005)
Delmonte, R., Bristot, A., Piccolino Boniforti, M.A., Tonelli, S.: Entailment and Anaphora
    Resolution in RTE3. In: ACL Workshop on Text Entailment and Paraphrasing, Prague,
    ACL Madison, USA, pp. 48–53 (2007)

Cui, H., Sun, R., Li, K., Kan, M.-Y., Chua, T.-S.: Question Answering Pas-sage Retrieval Using Dependency Relations. In: SIGIR 2005, pp. 400–406. ACM, Salvador (2005)

Litkowski, K.C.: Syntactic Clues and Lexical Resources in Question-Answering. In: Voorhees, E.M., Harman, D.K. (eds.) The Ninth Text Retrieval Conference (TREC-9), pp. 157–166. NIST Special Publication, Gaithersburg (2001)

Wang, M., Smith, N.A., Mitamura, T.: What is the Jeopardy Model? A Quasi-Synchronous Grammar for QA. In: Proceedings of the Conference on Empirical Methods in Natural Language Processing and Computational Natural Language Learning, Prague, pp. 22–32 (2007)

Traum, D., Habash, N.: Generation from Lexical Conceptual Structure. In: Workshop on Applied Interlinguas, ANLP 2000, Seattle, WA, pp. 123–134 (2003)

Opinion Mining and Sentiment Analysis Need Text Understanding

Rodolfo Delmonte and Vincenzo Pallotta

Abstract. We argue in this paper that in order to properly capture opinion and sentiment expressed in texts or dialogs any system needs a deep linguistic processing approach. As in other systems, we used ontology matching and concept search, based on standard lexical resources, but a natural language understanding system is still required to spot fundamental and pervasive linguistic phenomena. We implemented these additions to VENSES system and the results of the evaluation are compared to those reported in the state-of-the-art systems in sentiment analysis and opinion mining. We also provide a critical review of the current benchmark datasets as we realized that very often sentiment and opinion is not properly modeled.

1 Introduction

Sentiment analysis and Opinion mining [Pang and Lee 2008] are emerging applications of Natural Language Processing whose importance is becoming increasingly higher. Brand managers and market researchers use these applications in order to monitor the "voice of the customer" in order to study trends of acceptance/rejections and sometimes to discover issues in products or services.

We assume that in order to properly capture opinion and sentiment expressed in a text or dialog any system needs a full natural language understanding (NLU) approach. In particular, the idea that the task may be solved by the use of

Rodolfo Delmonte
Department of Language Science
Università "Ca Foscari"
Dorsoduro, 3462 - Venezia
30123 – Venezia, Italy
e-mail: delmont@unive.it

Vincenzo Pallotta
Department of Informatics
University of Fribourg
Bd. des Pérolles, 90,
1700 - Fribourg, Switzerland
e-mail: vincenzo.pallotta@unifr.ch

V. Pallotta, A. Soro, and E. Vargiu (Eds.): Advances in DART, SCI 361, pp. 81–95.
springerlink.com   © Springer-Verlag Berlin Heidelberg 2011

Information Retrieval tools like Bag of Words (BOWs) approaches has shown its intrinsic shortcomings. In fact, in order to achieve acceptable results, BOWs approaches are sometimes camouflaged by a keyword-based Ontology matching and Concept search, based on SentiWordNet[1] [Bentivogli et al. 2004], by simply stemming a text and using content words to match its entries and produce some result. Any search based on keywords and BOWs is fatally flawed by the impossibility to cope with such fundamental and pervasive linguistic phenomena as the following ones:

- presence of NEGATION at different levels of syntactic constituency;
- presence of LEXICALIZED NEGATION in the verb or in adverbs;
- presence of conditional, counterfactual subordinators;
- double negations with copulative verbs;
- presence of modals and other modality operators.

In order to cope with these linguistic elements we propose to build a Flat Logical Form (FLF) directly from a Dependency Structure representation of the content augmented by indices and where anaphora resolution has operated pronoun-antecedent substitutions. We implemented these additions our NLU system called VENSES [Delmonte et al. 2009]. The output of the system is an XML representation where each sentence of a text or dialog is associated to a list of attribute-value pairs, one of which is POLARITY. In order to produce this output, the system makes use of the FLF and a vector of semantic attributes associated to the verb at propositional level and memorized.

Important notions such as the distinction of the semantic content of each proposition into two separate categories, OBJECTIVE vs. SUBJECTIVE, are also required by the computation of opinion and sentiment. This distinction is obtained by searching for FACTIVITY markers again at propositional level. In particular we take into account:

- tense, voice, mood at verb level
- modality operators like intensifiers and diminishers, but also modal verbs
- modifiers and attributes adjuncts at sentence level
- lexical type of the verb (in Levin's classes and also using WordNet classification)
- subject's person (if 3rd or not).

The article is organized as follows. In section 2, we review the components of the VENSES system. In section 3 we present its tailoring for the task of sentiment analysis. Section 4 describes the experiment we carried out on a benchmark dataset where we compare our results to the state-of-the-are results and discuss the flaws of BOW systems. Section 5 concludes the paper with some lesson learned and recommendations.

---

[1] http://sentiwordnet.isti.cnr.it/

2   The VENSES System

VENSES is a tailored version of GETARUNS[2] [Delmonte 2007; 2008b], a complete system for text understanding developed at the Laboratory of Computational Linguistics of the University of Venice. The system produces different levels of analysis, from syntax to discourse. However, three of them contribute most to the success of sentiment analysis:

1. the syntactic and lexico-semantic module,
2. the anaphora resolution module [Delmonte et al. 2007],
3. the deep semantic module.

2.1   The Syntactic and Lexico-Semantic Module

GETARUNS, is organized as a pipeline which includes two versions of the system: what we call the *Deep* and *Partial* GETARUNS.

The Deep version of GETARUNS is equipped with three main modules: a *lower module* for parsing, where sentence strategies are implemented; a *middle module* for semantic interpretation and discourse model construction which is cast into Situation Semantics; and an *upper module* where reasoning and generation takes place.

GETARUNS, has a highly sophisticated linguistically based semantic module which is used to build up the DM. Semantic processing is strongly modularized and distributed amongst a number of different sub-modules, which take care of Spatio-Temporal Reasoning, Discourse Level Anaphora Resolution, and other subsidiary processes like Topic Hierarchy.

The architecture of the Partial GETARUNS is shown in Figure 1. This version is fired before the Deep system and is used as a back-off strategy whenever failures ensue. The Partial system tries at first to produce a full parse of the current utterance with the lower system. The Deep system makes use of chunks as produced by the Partial system, in order to guess where in the utterance is positioned the current analysis, in particular where the VP starts. Only in case the Deep system fails, the Partial system will proceed by producing Partial semantics and Discourse level analysis through middle and upper level.

The parser produces a c-structure representation by means of a cascade of augmented FSA[3]. Then it uses this output to map lexical information from a number of different lexica, which however contain similar information related to verb/adjective and noun sub-categorization. The mapping is done by splitting the

---

[2] The system has been tested in STEP competition (see [Delmonte 2008a], and can be downloaded in two separate places. The partial system called VENSES in its stand-alone version is available at:
   http://www.aclweb.org/aclwiki/index.php?title=Textual_Entailment_Resource_Pool
   The complete deep system is available both at:
   http://project.cgm.unive.it/html/sharedtask/

[3] Finite State Automata.

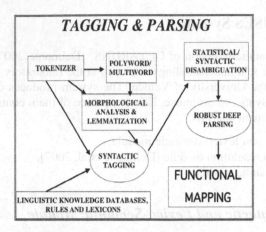

Fig. 1 GETARUNS: lower level.

sentences into clauses, which are normally main and subordinate clauses. Other clauses are computed in their embedded position and can be either complement or relative clauses.

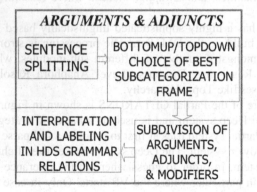

Fig. 2 GETARUNS: upper level.

The output of the upper level is what we call AHDS (Augmented Head Dependent Structure), which is a fully indexed logical form, with Grammatical Relations and Semantic Roles. The inventory of semantic roles we use is however very small (i.e. 35) even though it is partly overlapping with the set proposed in the first FrameNet project[4]. We prefer to use Generic Roles rather than specific Frame Elements (FEs) because sense disambiguation at this stage of computation may not be effective.

---

[4] http://framenet.icsi.berkeley.edu/book/book.pdf

2.2   The Anaphora Resolution Module

This module whose components are sketched in Figure 3 works on the so-called History List of entities present in the text so far. In order to make the output of this module usable by the Semantic Evaluator, we decided to produce a flat list of semantic vectors which contain all semantic related items of the current sentence. Inside these vectors, pronominal expressions are substituted by the heads of their antecedents.

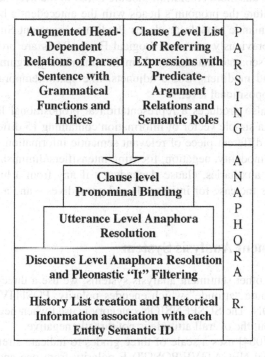

Fig. 3 GETARUNS: the anaphora module

The AHDS structure is passed to and used by a full-fledged module for pronominal and anaphora resolution, which is in turn split into *two sub-modules*. The resolution procedure takes care only of third person pronouns of all kinds – reciprocals, reflexives, possessive and personal. Its mechanisms are quite complex and details can be found in [Delmonte et al. 2006]. The *first sub-module* basically treats all pronouns at sentence level – that is, taking into account their position – and if they are left free, they receive the annotation "external". If they are bound, they are associated to an antecedent's index; else they might also be interpreted as expletives, i.e. they receive a label that prevents the following sub-module to consider them for further computation.

The *second sub-module* receives as input the external pronouns, and tries to find an antecedent in the previous stretch of text or discourse. To do that, the system computes a *topic hierarchy* that is built following suggestions by [Grosz and Sidner 1986] and is used in a centering-like manner.

2.3 The Deep Semantic Module

The output of the anaphora resolution module is fed to the deep semantic module in order to substitute the pronoun's heads with the antecedent's heads. After this operation, the semantic module produces Predicate-Argument Structures (PASs) on the basis of previously produced Logical Form. PASs are produced for each clause and they separate mandatory from non-mandatory arguments, and these from adjuncts and modifiers. Some adjuncts, like spatio-temporal locations, are only bound at propositional level.

This module also produces a representation at propositional level, which for simplicity is just a simple vector of information containing 15 different slots, each one containing a different piece of relevant semantic information. We encode the following items: modality, negation, focusing intensifiers/diminishers, manner adjuncts, diathesis, auxiliaries, clause dependency if any from a higher governing predicate – this is the case for infinitivals and gerundives – and a subordinator, if any.

3 The Sentiment Analysis System

Differently from other sentiment analysis systems, we use a three-way classification for the attribute "attitude" which encodes polarity: POSITIVE, NEGATIVE and SUSPENSION. The SUSPENSION category is used when negation is present in the utterance but the overall attitude is not directly negative.

[Hu and Liu 2004] uses a scale of three grades to indicate strength and distinguish cases of real NEGATIVE/POSITIVE polarity from one another. In the readme file associated to the datasets, the authors comment on this grading system that: *"… note that the strength is quite subjective. You may want to ignore it, but only considering + and –"*.

It is a fact that annotation criteria are hard to establish, but then the outcome will always subjective in a sense. For instance in the example below taken from one of the datasets made available by the authors and on which we evaluated our system, the score [-1] indicates low negative polarity strength.

In order to have an idea of where the problems lie, we report below how the example was annotated in the dataset (A), and then the output of our system (B):

A. viewfinder[-1]##the lens is visible in the viewfinder when the lens is set to the wide angle , but since i use the lcd most of the time , this is not really much of a bother to me.
B. id="44", predicate="be", topic="lens", attitude="suspension" factivity= "factive_statement"

It is apparent that this example cannot be considered as fully negative and probably the negative category alone does not capture the nuances of the statement. We take a conservative approach and we classify this statement as being SUSPENSION because negative expressions cannot be directly mapped onto a negative opinion. Here and elsewhere we annotated SUSPENSION, and the system correctly labels the example: this label indicates an attitude which is not strongly marked for either polarity value, and in some cases this may also be due to the presence of double negation. As a consequence, we also use SUSPENSION in the following example, where (Hu and Liu 2004) annotates instead as Positive with high confidence:

A. weight[+2]##at 8 ounces it is pretty light but not as light as the ipod .
B. id="46", predicate="be", topic="light", attitude="suspension" factivity= "factive_statement"

It is apparent that this case cannot be computed as a strong case for positive attitude, actually this is not a positive opinion at all. We would like to stress here that those shown in A are manual annotations and not output from an automatic classification system. This also entails that the training dataset is intrinsically flawed due to a misunderstanding of what polarity actually is.

In many other cases, Hu and Liu's provide no annotation, which does not mean, in our opinion, that the utterance can be considered as neutral, as for instance in the following cases:

A. ##if you have any doubts about this player , well do n't .
B. id="32",predicate="have" topic="player" attitude="suspension" factivity= "opinion_internal"

A. ##can 't complain and i recommend it over all the other players , just hope that remote will come out soon .
B. id="27", predicate="recommend", topic="player", attitude="suspension", factivity= "opinion/ factive_statement"

In many cases, however, it is hard to understand the reason why the annotation has not been made available for trivial cases as for instance in,

- "do not buy this player"
- "a piece of junk"
- "don't waste your money".

3.1 Sentiment Analysis of Conversations

We tailored our system to deal with contexts larger than short reviews of products. Here below we present an excerpt from a short dialog, which contains a certain number of complex negative cases to solve. In Table 1 we provide our analysis of each sentence:

Table 1 Sentiment and Factivity analysis of a conversation excerpt.

id	Content	Predicate	Polarity	Factivity
1	Well, what do you think?	say	positive	question
2	That's not so bad.	be	suspension	opinion_internal
3	I'm not complaining.	complain	suspension	opinion_statement
4	That's not true.	be	negative	opinion_statement
5	Jack never contradicts my opinions.	contradict	suspension	factive_statement
6	Mark always contradicted my ideas.	contradict	negative	factive_statement
7	Mark never accepted disadvantages.	accept	negative	factive_statement
8	Nobody bought that product.	buy	negative	factive_statement
9	I bought an awful product.	buy	negative	factive_statement
10	I don't like that product.	like	negative	factive_statement
11	I strongly criticize such a product.	criticize	negative	factive_statement
12	No sensible customer would buy that product.	buy	negative	factive_statement
13	Mary bought that product for an awful purpose.	buy	negative	factive_statement
14	Mary bought that product to kill herself.	buy	negative	factive_statement
15	Mark didn't make a bad deal.	make	relevant	opinion_statement
16	That product doesn't seem to be awful.	seem	suspension	opinion_internal
17	Mary didn't buy that awful product.	buy	negative	factive_statement
18	John didn't kill the bad feelings of the customers about that awful product.	kill	suspension	factive_statement

As it can be easily noticed, the problem is due to the presence of negation that is not solvable by a simple one-way decision – yes/no. In many cases the information about the attitude of the speaker is just not directly communicated and needs further specification. In sentence 16, for instance, is not a straightforward admission of disagreement; the same applies to sentence 18. We also regard sentences 2, 3, 5 to be cases of *indirect judgment,* which however is not explicit enough to be assigned to a positive attitude. For this reason, we decided to introduce the SUSPENSION marker, which encodes all cases of indirect judgment, and other similar situations.

Coming now to clear cases of NEGATIVE attitude, we register sentences like 4, 6, 7, 8, 9, 10, 11, 12, 13, 14 and 17. However, not all these sentences can be easily understood as being totally NEGATIVE. In particular, only sentence 4 and 10 are simple cases of negation at main verb level and may be computed safely as cases of negative attitude. Sentences 6 and 11 are again cases of negative attitude but there is no explicit negation expressed: just negatively marked verb at lexical level.

Examples 8 and 12 express negation at subject level: as can be gathered, this can only be evaluated as a real negative attitude only in case the main verb indicates positive actions. Apparently, these cases can also be contradicted by the same speaker, by using BUT and other adversative discourse markers (e.g. "even though nobody likes it..."; "nobody likes it", "but ..."). Examples 9, 13, and 14 introduce negative attributes at object and complement level. This is also computed by the system as a case of negative attitude.

The system also computes as negative example 17, which is a case of double negation: in this sentence, negation is present both at verbal level and at complement level. This might be understood as positive attitude (i.e. "if she did not do that then it is good...."). However we assume that this is also interpretable as a report of something negative that might have happened and not as a negative judgment. This distinction may seem subtle, but we believe is very important in order to avoid false positives in the classification. Eventually, we have also cases of SUSPENSION involving the presence of negation as example 18 shows.

An important subdivision of all semantic types involved, regards FACTIVITY, which, as we said before, can constitute an important indicator of the speaker's attitude in uttering a given judgment. It is important to mention that sentences are in fact utterances that can be categorized in at least two main types:

1. they constitute OBJECTIVE (or FACTIVE) STATEMENT reporting in this way some fact usually in third person subject;
2. they may constitute SUBJECTIVE (or NON-FACTIVE) OPINIONs expressed by the speaker him/herself in first person or reporting on somebody elses's opinion.

Opinions are always subjective but may also report an internal thought, a wish, a hope, or else a definite state, event, and activity by the subject. In the former case, we use OPINION_INTERNAL, to highlight the weight of subjective markers as in 2 and 16. In the latter case, we use OPINION_STATEMENT because it is either the case that the utterance refers acts or events of third persons, as in 15; or else, it reports the evaluation of the speaker as in 3 and 4. Other markers are QUESTION, which can still be computed as either positive or negative; and RELEVANT, implying some indirect judgment as shown by 16 and double negation and reinforcing on SUSPENSION.

As a last example, we now consider a really difficult utterance to evaluate from Hu and Liu's dataset:

*Positive-1 dvd - so far the dvd works so i hope it doesn't break down like the reviews i 've read.

This item has been correctly annotated as POSITIVE in Hu and Liu's dataset. However, in order to capture the dependency between the negated sentence and the "hope" predicate, a system definitely needs to build a logical form and all the appropriate indices. Predicates like "hope" make the following governed proposition "opaque" and non-factive. This forces the system to "dummify" the presence of negation. Looking carefully at the example, this sentence cannot be considered as positive as it is a clear case of SUSPENSION where no judgment has been expressed, but only a worry that other reviews made would turn out to be true in reality.

3.2   The Semantic Markers: CONDITIONAL and COMPARATIVES

Eventually, there are important components of a semantic analysis, which may heavily influence the final output. We are am now referring to two well-known cases discussed in the literature: the presence of "conditional" discourse markers like IF, WHETHER which transform a statement into a conditional clause which is usually accompanied by the presence of "unreal" mood like conditional or subjunctive. And then we come to "comparative" constructions, which are more frequent in consumer product reviews than in blogs or social networks opinions. As far as comparatives are concerned, it is a fact that real utterances contain a gamut of usage of such a construction, which is very hard to come to terms with. We list some of the most relevant cases here below and then make some comments. Each utterance is taken from Hu and Liu's reviews databases and has an evaluation at the beginning of the line:

> a. *Positive-2 player - i did not want to have high expectations for this apex player because of the price but it is definitely working out much better than what i would expect from an expensive high-end player.
> b. *Positive-2 look - without a doubt the finest looking apex dvd player that i 've seen.
> c. *Positive-2 dvd player - so sit back , relax and brag to all your friends who paid a mountain of money for a dvd player that can't do half the things this one can , and for a fraction of the price !
> d. *Positive-3 camera - recent price drops have made the g3 the best bargain in digital cameras currently available.
> e. *Positive-2 feel - you feel like you are holding something of substance , not some cheap plastic toy.
> f. *Positive-3 camera - i can't write enough positive things about this great little camera !
> g. *Positive-3 camera - this is my first digital camera and i couldn't be happier.
> h. *Positive-3 finish - its silver magnesium finish is stunning, and the sharp lines and excellent grip are better than any other camera i've seen.
> i. *Positive-2 noise another good thing is that this camera seems to introduce much less noise in dark places than others i've seen.
> k. *Positive-2 camera this is by far the finest camera in its price and category i have ever used.

As it can be noticed, in many cases what is really the guiding principle is the need of comparing the evaluative content of two opposing propositions, rather than simply measuring degree of comparison (i.e. superlative rather than comparative grade). In example a. the first proposition is negated and then the second compared proposition marked by BUT is a really hard complex sentence to compute. In b. one has to compute correctly "without a doubt". In c. the first proposition has a relative clause referring to a negative fact, where however the governing verb BRAG can be understood both negatively and positively. In d. the phrase "recent price drops" can be a negative fact but has to be understood positively together with the following proposition where "best bargain" appears. Again in e. one needs to compare two propositions one of which has an ellipsed VP. In f. the reviewer uses a rhetorical device "can't write enough positive..." which however introduces negation. The same applies to example g.

4   The Experiment with Products Reviews

In order to evaluate our system, we used [Hu and Liu 2004] datasets, which have been collected and partially annotated in 2004 and are made of customer reviews of 5 products downloaded from Amazon.com. In fact, we used for perusal and evaluation only three of them: Canon (digital camera), Creative (mp3 player) and Apex (dvd player). The problem was that the annotated examples were just a small percentage of the total - 1302 sentences over 3300, so we had to manually annotate the remaining cases (60% of all utterances) ourselves and make some corrections on the input: the texts were full of typos and had many non-words, fragments, ungrammatical sentences etc. Overall, we parsed 30,000 tokens and 3300 utterances. In Table 2 we report some statistics about the three datasets we have used in our experiment. The Sents column indicates the number of total utterances present in each dataset.

Table 2 Annotation data from Ho and Liu's datasets

	Positive	Negative	Totals	Sents
apex	148	195	343	840
canon	184	54	238	643
creative	421	299	720	1811
totals	753	548	1301	3394

As can be easily noticed, only 38.33% (1301 out of 3394) of all utterances have been annotated, which makes the comparison fairly difficult to make. In particular, if we look at our annotated data in Table 3, the overall number of NEGATIVE polarity judgments constitutes 58% of all judgments when compared to 42% in Ho and Liu's annotations. The final outcome is then totally mistaken: in our case the judgments are more negative and in Ho and Liu's they are more positive disregarding the subdivision of reviews into each separate products. We computed the number of annotations in original datasets, which have been graded [+/- 1], thus indicating that the confidence of the annotator is very low, and this makes up 16.37% of all annotations. In our case, the SUSPENSION annotations constitute 23.22% of all annotations.

Table 3 Automatic annotation with VENSES

	Pos.	Neg.	Susp.	Quest.	Totals
apex	327	300	199	15	841
canon	294	197	143	13	647
creative	558	782	430	37	1797
totals	1030	1447	769	65	3311

The first interesting fact to notice is the slight difference in Recall, where we see that of all the utterances present we only got 97.55%. It is important to highlight the difference in the approach. Our system's output refers to *real utterances*, which sometimes do not coincide with each line or record in the input file. The system computes an utterance every time it finds a sentence delimiting punctuation mark. As a result, in some cases, as in "canon" dataset, we end up with additional utterances to evaluate.

Table 4 Evaluation on the basis of Hu and Liu's gold standard

	Accuracy	Accuracy %	F-score
apex	286/343	83.39	90.94%
canon	174/238	73.00	84.39%
creative	547/720	76.20	86.49%
totals	1007/1301	77.40	87.26%

The results of the evaluation shown in Table 4 are based on Hu and Liu's dataset at first and are computed for accuracy as a ratio of correct/gold standard; we also compute the F-score[5], where Recall is in our case equal to 100% in the sense that we compute all evaluations for all sentences.

Table 5 Evaluation on the basis of our annotation

	Precision	Recall	F-score
apex	670/841=79.66%	826=98.41%	88.05%
canon	468/648=72.23%	638=98.45%	83.32%
creative	1424/1811=78.63%	1764=97.40%	87.01%
totals	2562/3300=77.60%	3228/3300=97.81%	86.54%

Table 5 shows results computed on the basis of the overall annotation integrated by our single annotator, has a slightly lower F-score. It is interesting to note the difference in overall polarity evaluation, which may affect the opinion of prospective customers inducing them in buying or not buying a certain product on the basis of the balance between positive and negative polarity. In the evaluation carried out by Ho and Liu we see that negative judgments constitute the 42.12% (548 negative and 753 positive). In our case the proportions are reverse: we have (1030 positive and 1447 negative) 58.42% negative judgments.

Data related to SUBJECTIVITY and FACTIVITY reported in Table 6 show a balanced subdivision of all data between the two categories.

---

[5] The F-score is 2*precision*recall/(precision+recall).

Table 6. Subjectivity results from VENSES

	Fact	Opin	Opin_ Inter	Fact/Opin	Total
apex	398	265	87	100	850
canon	300	226	47	75	648
creative	772	609	195	219	1795
totals	1470	1100	329	394	3293

4.1 The Experiment with Quotations

We did another experiment on the basis of a corpus of news annotated at the JRC of the European Commission [Balahur et al. 2010]. The corpus is a collection of 1592 quotations, which however have been collected automatically. It comes out that they contain 151 totally repeated texts, 13 non-sentences or even non-fragments that cannot be evaluated at all. Then there are some 24 quotes which are constituted by questions, which again not being statements cannot be evaluated negatively/positively. Another 15 quotations are portions of quotes and have been included in the evaluation. At the end we came up with 1404 quotations. In fact, as the authors report in their paper, only 1292 are fully agreed quotes on the basis of their three annotators. The evaluation the authors present at the end of their paper is based on a small amount of data - 427 quotes - constituted only by those quotes, which have received full agreement in their polarity judgments. The authors leave out 865 quotes, which have been computed as objective statements, on the assumption that only subjective quotes can be regarded appropriate for polarity judgment.

We don't agree at all with their definition of polarity judgment: "statements may describe objective negative state of affairs, current or even future events much in the same way in which subjective statements do". In fact, since quotations are mainly third person descriptions, narrations, or reported speech they can belong to both categories.

The results in terms of accuracy are similar to those on reviews and dialogs: 82.05% for negative judgments, 70.34% for positive judgments, overall 76.2% accuracy. The main difference is constituted by the evaluation of positive judgments, which are indirectly reported and require a lot of semantic knowledge. Consider quotes like the following ones that are evaluated for positive:

1. "anybody who wants [Mr Obama] to fail is an idiot, because it means we're all in trouble... ".
2. "Charles Freeman was the wrong guy for this position. His statements against Israel were way over the top and severely out of step with the administration. I repeatedly urged the White House to reject him, and I am glad they did the right thing."

The first sentence is correctly classified as non-negative with respect to the main topic, Obama, as well as for the second, which is a reply to a negative comment and it should thus count as positive. These cases are clearly hard to capture for a system without a deep language understanding and that is capable to deal with larger context than a single unit.

5 Conclusions

In this paper we have advocated the need of a Natural Language Understanding system to adequately deal with the task of sentiment analysis and opinion mining. We pointed out the issues in both annotated datasets used for benchmarking and the mainstream methods that are based on enhanced Bag of Word approaches. Our conclusions that Sentiment Analysis and Opinion Mining community needs better datasets and more precise annotation guidelines.

For what concern the performance of our system, we found that not being based on Machine-Learning, it can be substantially improved on the basis of better rules and better lexica. In fact, most mistakes are due to the presence of wrong polarity assignments in the lexical resources used such as the Harvard's General Inquirer dictionary[6] or the Wiebe's list [Wiebe and Mihalcea 2006]. In fact, what can apply to psychology tests does not always apply to the evaluation of reviews, which have products as their objects. In addition, we discovered that there are a variable number of cue words that are ambiguous and vary their connotation (i.e. from positive may become negative and vice versa) according the domain of application. In the field of photography for instance, words such as shoot or shot do not carry negative connotation. So eventually, the system must be updated with respect to the domain and this is something that can be done using WordNet Domains, a resource made freely available by IRST/FBK (see [Bentivogli et al. 2004]), which indicates domains with the needed perspicuity.

References

[Balahur et al. 2010] Balahur, A., Steinberger, R., Kabadjov, M., Zavarella, V., van der Goot, E., Halkia, M., Pouliquen, B., Belyaeva, J.: Sentiment Analysis in the News. In: Proceedings of the 7th International Conference on Language Resources and Evaluation (LREC 2010), Valletta, Malta, May 19-21, pp. 2216–2220 (2010)

[Bentivogli et al. 2004] Bentivogli, L., Forner, P., Magnini, B., Pianta, E.: Revising the WORDNET DOMAINS Hierarchy: semantics, coverage and balancing. In: Proceedings of COLING 2004 Workshop on Multilingual Linguistic Resources, Geneva, Switzerland, August 28, pp. 101–108 (2004)

[Delmonte 2007] Delmonte, R.: Computational Linguistic Text Processing – Logical Form, Logical Form, Semantic Interpretation, Discourse Relations and Question Answering. Nova Science Publishers, New York (2007)

[Delmonte et al. 2007] Delmonte, R., et al.: Another Evaluation of Anaphora Resolution Algorithms and a Comparison with GETARUNS' Knowledge Rich Approach. In: Proceedings of ROMAND 2006 - 11th EACL, Geneva, pp. 3–10 (2006)

---

[6] http://www.wjh.harvard.edu/~inquirer/homecat.htm

[Delmonte 2008a] Delmonte, R.: Computational Linguistic Text Processing – Lexicon, Grammar, Parsing and Anaphora Resolution. Nova Science Publishers, New York (2008)

[Delmonte 2008b] Delmonte, R.: Semantic and Pragmatic Computing with GETARUNS. In: Bos, Delmonte (eds.) Proceedings of Semantics in Text Processing (STEP), Research in Computational Semantics, vol. 1, pp. 287–298. College Publications, London (2008)

[Delmonte et al. 2009] Delmonte, R., Tonelli, S., Tripodi, R.: Semantic Processing for Text Entailment with VENSES. In: Proceedings of the TAC 2009 Workshop on TE, Gaithersburg, Maryland (November 17, 2009)

[Hu and Liu 2004] Hu, M., Liu, B.: Mining and summarizing customer reviews. In: Proceedings of KDD 2004 (2004)

[Pang and Lee 2008] Pang, B., Lee, L.: Opinion mining and sentiment analysis. Foundations and Trends in Information Retrieval 2(1-2) (2008)

[Grosz and Sidner 1986] Grosz, B., Sidner, C.: Attention, intentions, and the structure of discourse. Computational Linguistics 12(3), 175–204 (1986)

[Wiebe and Mihalcea 2006] Wiebe, J., Mihalcea, R.: Word Sense and Subjectivity. In: Proceedings of ACL 2006 (2006)

Sentiment Analysis of French Movie Reviews

Hatem Ghorbel and David Jacot

Abstract. In sentiment analysis of reviews we focus on classifying the polarity (positive, negative) of conveyed opinions from the perspective of textual evidence. Most of the work in the field has been intensively applied on the English language and only few experiments have explored other languages. In this paper, we present a supervised classification of French movie reviews where sentiment analysis is based on some shallow linguistic features such as POS tagging, chunking and simple negation forms. In order to improve classification, we extracted word semantic orientation from the lexical resource SentiWordNet. Since SentiWordNet is an English resource, we apply a word-translation from French to English before polarity extraction. Our approach is evaluated using French movie reviews. Obtained results showed that shallow linguistic features has significantly improved the classification performance with respect to the bag of words baseline.

1 Introduction

Sentiment analysis is an emerging discipline whose goal is to analyze textual content from the perspective of the opinions and viewpoints they hold. A large number of studies have focused on the task of defining the polarity of a document which is by far considered to be a classification problem: decide to which class a document should be attributed; class of positive or negative polarity.

Most of the work in the field has been intensively applied on the English language. For this purpose, English corpora and resources (such as MPQL [WWH05], Movie Review Data [PLV02], SentiWordNet [ES06] and WordNet-Affect [SV04]) have been constructed to aid in the process of automatic supervised and unsupervised polarity classification of textual data. Nevertheless, there remain very few experiments applied to other languages.

Hatem Ghorbel · David Jacot
University of Applied Sciences Western Switzerland, Haute Ecole Arc Ingénierie,
St-Imier, Switzerland
e-mail: hatem.ghorbel@he-arc.ch, david.jacot@master.hes-so.ch

V. Pallotta, A. Soro, and E. Vargiu (Eds.): Advances in DART, SCI 361, pp. 97–108.

In this context, we address in this paper the issue of polarity classification applied to French movie reviews. We used a supervised learning approach where we trained the classifier with annotated data of French movie reviews extracted from the web. As classification features, beyond the word unigrams feature taken as the baseline in our experiments, we extracted further linguistic features including lemmatized unigrams, POS tags, simple negation forms and semantic orientation of selected POS tags. The latter is extracted from the English lexical resource SentiWordNet after applying a word-translation from the French to English.

The main goal of our experiments is firstly to confirm that the incorporation of shallow linguistic features into the polarity classification task could significantly improve the results. Secondly, to address the problem of loss of precision in defining the semantic orientation of word unigrams from English lexical resources, mainly due to the intermediate process of word-translation from French to English correlated with further issues such as sense disambiguation.

In the rest of the paper, we commence by briefly describing the previous work in the field of sentiment analysis and polarity classification. Then we describe the set of extracted features used in polarity classification of French movie reviews. Finally we provide and discuss the obtained experiment results and end up by drawing some conclusions and ideas for future work.

2   Previous Work

Classical approaches in text retrieval and categorization has so far focused on mining and analyzing factual information such as entities, events and their properties. They basically utilize Natural Language Processing methods and techniques in order to extract objective features aiding the classification and categorization of textual expressions [PLV02] with special emphasis on linguistic features in order to increase the performance. As linguistic features [Gam04, MTO05] present syntactically motivated features, most of them based on dependency path information and modeled as high n-grams. Further linguistic features such as part of speech, forms of negation, verbs modality, and semantic information (from WordNet for instance) are recently explored [WK09, TNKS09].

Much of the previous work focuses on defining the characteristics of conveyed opinions on the basis of textual data with processing granularity ranging from words, to expressions, sentences and documents. At the word level, considered as the most fine-grained level, it is assumed that there is a one-sided opinion throughout the whole sentence or the expression, typically the case of opinion question answering or human discussions. We mainly discern two types of research approaches that aim at solving this problem: statistical and semantic approaches. Statistical approaches make use of learning techniques to classify the semantic polarity of conveyed opinion into positive and negative classes and approximate the value of their intensity. These techniques vary from supervised to unsupervised learning, typically probabilistic methods (such as Naive Bayes, Maximum Entropy), linear discrimination (such as Support Vector machine) and non-parametric classifiers (such as K-Nearest Neighborhood) as well as similarity scores methods (such as phrase pattern matching, distance vector, frequency counts and statistical weight measures). For this purpose, a large number of annotated corpora and sentiment oriented resources have been constructed to support supervised learning like for instance MPQL [WWH05], Movie Review Data [PLV02], SentiWordNet [ES06], WordNet-Affect [SV04]), Product Review [YNBN03], Book Review [GA05], the Whissell's Dictionary of Affect Language [Whi89], Linguistic Inquiry and Word Count Dictionary (LIWC2001)[JP01], in addtion to the huge amount of available raw sentiment oriented data found in forums, blogs, chat rooms, review, debates and E-opinion web sites .

In order to support learning features mostly based on lexical and shallow linguistic attributes (such as lemmas, POS tags, simple negation forms), statistical approaches are substantially coupled with semantic approaches in order to achieve better results [KH04, WWH05]. Generally, semantic approaches improves sentiment classification by integrating features from common sense ontologies, sentiment and lexical-semantic resources. For instance, [HL04, ES05, NSS07, Den08] classify polarity using emotion words and semantic relations from WordNet, WordNet Gloss, WordNet-Affect and SentiWordNet respectively.

The word level is considered as the basic level and the background of the research work in the field. One of the main issues still to be resolved is how to define and measure the semantic orientation of a word or an expression in its context and whether the use of such a measure would improve polarity classification. Some studies showed that restricting features to those adjectives would improve classification performance. [HM97] have focused on defining the polarity of adjectives using indirect information collected from a large corpus. They take advantage of linguistic constraints such as parallel structures and conjunctions (for example nice and comfortable, either or, neither nor, but, or, and, etc.) and morphological relationship (for example helpful, helpless) to elaborate a kind of link prediction between adjectives ultimately used to cluster them into positive and negative clusters. However, more research showed that most of the adjectives and adverbs, as well as small groups of nouns and verbs possess semantic orientation [TTC09, AB06, ES05, GA05, TIO05, TL03]. Automatic methods of sentiment annotation at the word level can be grouped into two major categories: (1) corpus-based approaches and (2) dictionary-based approaches.

2.1 Corpus-Based Approaches

The first category includes methods that rely on statistical measures between patterns of words in large corpora to determine their sentiment orientation [HM97]. For instance, [TL02] determined the statistical similarity between two words by counting the number of results returned by web searches joining the words with a NEAR operator (Altavista). Further methods based on the use of seeds (reference terms with already known polarity such as *good* and *poor* also called sentiment-bearing terms) have been used by [Tur02] who measures the mutual information between the phrase and the positive word *excellent* minus the mutual information between the phrase and the negative word *poor*. [YH03] use the same co-occurrence statistics but applying the Modified Likelihood ratio to measure the similarity between words and positive/ negative seeds. The similarity between a candidate word and a set of manually selected seed words was used to place it into a positive or negative polarity. [HR10] applied a Markov random walk model to a large word relatedness graph, producing a polarity estimate for any given word. Their experiment could be used both in a semi-supervised settings where a training set of labeled words is used, and in an unsupervised setting where a handful of seeds is used to define the two polarity classes.

2.2 Dictionary-Based Approaches

The second category includes the use of WordNet and other dictionary information, especially, synsets, hierarchies and semantic relations. [HL04] approach is based on acquiring synonyms and antonyms of a set of seed sentiment words in Word-Net. Similarly [KH04] use WordNet to obtain a synonym set of the unseen word to determine how it interacts with the elaborated sentiment seed lists using Naive Bayes method. [JKdR04] use WordNet to construct a network by connecting pairs of synonymous words. The semantic orientation of a word is decided by its shortest paths to two seed words *good* and *bad* representing positive and negative orientations. [ES05] determine the orientation of words by analyzing glosses in an online glossary or dictionary. They have used WordNet Gloss and WordNet as the source of semantic relations such as synonymy, antinomy and hyponymy to expand lexical information in sentences before performing statistic classification. Their classifier is trained on glosses of selected seed words and is then applied to classify gloss of an unseen word. Similarly, [TIO05] construct a lexical network by connecting similar/related words where each node has a semantic orientation value and the neighboring nodes related with weighted links tend to have the similar values. The orientation value is estimated using the probabilistic network Potts Model [YH82] constructued from gloss information in dictionaries.

When it comes to target languages, most of the previous work have focused on English and Asian languages (mostly Japanese and Chinese). We find however few works [Den08, ACS08] that have explored sentiment analysis in a multilingual framework such as Arabic, Chinese, English, German and Japanese. Their methodology is based on standard translation from target language to English in order to reuse existing English corpora and resources for polarity classification.

3 Feature Design

Similarly to previous sentiment analysis studies, we have defined three categories of features. These include lexical, morpho-syntactic and semantic (word polarity) features. Lexical and morpho-syntactic features have been formulated at the word level, whereas semantic features have been formulated at the review level.

3.1 Lexical Features

This is the baseline of our experiments and is mainly composed of word unigrams. The global assumption in this choice is that we tend to find certain words in positive reviews and others in negative ones. Each unigram feature formulates a binary value indicating the presence or the absence of the corresponding word at the review level.

Lemmatization is argued to be relevant in sentiment analysis in order to group all inflected forms of a word in a single term feature. For example the words *aimé*, *aimait* and *aimer* share the same polarity but will be considered as five separate features during the classification. When applying lemmatization, we would obtain a unique feature. Features reduction would improve the tuning of the training process.

3.2 Morpho-Syntactic Features

Some studies showed that restricting features to specific part-of-speech (POS) categories would improve performance (for instance [HM97] have restricted features to adjectives). In our approach, POS tags are proposed to be used to enrich unigrams features with morpho-syntactic information so as to disambiguate words that share the same spelling but not the same polarity. For example, it would distinguish the different usages of the word *négatif* that can either be a neutral noun *un négatif* or a negative adjective *un commentaire négatif*. Moreover POS tags are useful to handle negation and to aid word sense disambiguation before polarity extraction in SentiWordNet as it will be detailed hereafter.

Negation is handled at the shallow level of morpho-syntactic constituency of sentences avoiding the heavy processing of its deep syntactic structure. The detection of negated forms is performed by searching specific patterns formed from the abundantly utilized lexicalized forms of negation combined with particular n-grams of POS categories. We defined two simple patterns that cope with the negation form (1) at the verb level for example *le scénario ne brille pas* and (2) at the adjective and noun level for example *sans histoire originale*.

The scope of the negation is fixed with respect to a predefined context window of *n* POS categories within a textual span limited by a punctuation sign. We invert the polarity of the *n* verbs, nouns and adjectives within the context of each detected negation. We do not cope with other composed forms of negation such as conditional, double negation, the counter factual subordinates and modalities. The entailments of such a polarity inversion are first situated at the lexical level; unigrams features are inverted during features vector construction that is if we consider the previous example, in stead of having in the feature *original*, we would have a separate feature *!original* in the vector; second at the semantic level, polarity is inverted from positive to negative and vice-versa in the calculation of the overall polarity of a review as we will detail in the following section.

3.3 Semantic Features

As argued in previous works [HL04, ES05, NSS07, Den08], the incorporation of corpus and dictionary based resources such as WordNetAffect, SentiWordNet and

Whissell's Dictionary of Affect Language contributes in improving the sentiment classification. Based on such results, we use the lexical resource SentiWordNet[1] to extract word polarity and calculate the overall polarity score of the review for each POS tag. SentiWordNet is a corpus-based lexical ressource constructed from the perspective of WordNet. It focuses on describing sentiment attributes of lexical entries describe by their POS tag and assigns to each synset of WordNet three sentiment scores: positivity, negativity and objectivity.

Since SentiWordNet describes English lexical resources, we go through a word-translation from French to English before polarity extraction. Words are lemmatized before being passed through the bilingual dictionary. We use POS information as well as the most frequently[2] used sense selection to disambiguate senses and predict the right synset. We only considered the positivity and the negativity features for the four POS tags noun, adjective, verb and adverb for this task.

More specifically, to each review we added two features holding the overall measures of negativity and positivity of words described with a ADJ, ADV, NOUN and VERB POS tags. These measures are extracted from SentiWordNet and defined as the sum of polarities over all the words of the review respecting POS categorization. For example, for a given review, we obtain the following semantic features vector *(neg_adv, 6.38); (pos_adv, 1.25); (neg_noun, 0.12); (pos_noun, 0.50); (neg_adj, 0.62); (pos_adj, 0.12); (neg_verb, 0.12); (pos_verb, 0.38).*

4   Experiments

Since we didn't find any available sets of annotated data (already classified as negative or positive) of French movie reviews, we collected our own data from the web[3]. We extracted a corpus of 2000 French movie reviews, 1000 positive and 1000 negative, from 10 movies, 1600 were used for training and 400 for testing. We included reviews having a size between 500 and 1000 characters.

Prior classification of the corpus is elaborated according to user scoring: positive reviews are marked between 2.5 and 4 whereas negative reviews are marked between 0 and 1.5[4]. This prior classification is based on the assumption that the scoring is correlated to the sentiment of the review.

For our experiments, the data was preprocessed with the TreeTagger[Sch94], a French POS tagger and lemmatization tool. We applied Support Vector Machine (SVM) classification method and utilized SVM^{Light} [Joa98] classification tool with its standard configuration (linear kernel) to implement a series of experiments where each time we define a set of combined features and evaluate the accuracy of the approach.

---

[1] SentiWordNet 1.0.1

[2] This choice is based on the assumption that reviewers spontaneously use an everyday language.

[3] We extracted spectators reviews from http://www.allocine.com

[4] Scores are bounded between 0 (for very bad) and 4 (excellent) with a step of 0.5. Reviews scored with 2 are not considered in the construction of our corpus since it is hard to manually classify them as positive or negative opinions.

Table 1 Performance of most relevant feature sets.

Features	# of features	Results [%]		
		Pos.	Neg.	Global
(1) Unigrams	14635	92.00	91.00	91.50
(2) Unigrams + lemmatization	10624	92.00	93.00	92.50
(3) **Unigrams + lemmatization + negation**	**12002**	**92.50**	**94.00**	**93.25**
(4) Unigrams + lemmtization. + POS	12229	93.00	92.50	92.75
(5) Unigrams + lemmatization + POS + negation	13625	92.50	93.50	93.00
(6) Unigrams + lemmatization + POS (ADJ)	2109	79.50	92.00	85.75
(7) Unigrams + lemmatization + POS (ADJ) + negation	2492	80.00	91.00	85.50
(8) **Unigrams + lemmatization + polarity**	**10632**	**93.00**	**93.50**	**93.25**
(9) Unigrams + lemmatization + negation + polarity	12010	93.00	92.50	92.75

4.1   Results and Discussion

The results of the following experiments are summarized in Table 1 above. For each experiment labeled from (1) to (9), we present the number of used features and the accuracy measured on the test corpus.

4.1.1   Lexical Features

Similarly to [PLV02] we encoded all words features as binary values indicating the presence or the absence of a word in a review. As a first step, we included the entire set of words without applying any specific filtration method.

The accuracy in experiment (1) using the entire set of words is found 91.50%; when comparing this result to Pang et al. [PLV02] who reported an accuracy of 82.90% on English movie reviews using similar features, we find that our results are approximately 10% higher. We believe that this gap is due to the nature of our corpus and the size of our reviews (the collected French reviews are shorter). Moreover, the incorporation of the lemmatization process (2) increases the accuracy by 1.00% up to 92.50%. This was quite expected since French is an inflected language. In experiment (3) we find that negation, although it is processed in a simple form, improves the results to reach 93.25%. Moreover, we notice that the classification of the negative reviews is being improved by the negation processing (from 93% up to 94%) which noticeably means that negation is relatively efficient at this lexical level.

After looking deeply through the reviews, we found that misclassification is mainly due to the following difficulties.

Misspellings. Misspelled words are not standard unigrams and hence could not regularly be present in the training data. Reviews containing a large number of misspellings would have their features significantly reduced and so provide very poor information for the classification. We noted that isolated and common misspellings don't affect much the classification but reviews which contain a relatively large number of misspellings tend to be misclassified. Sometimes misspellings are hard to be automatically corrected, especially those made voluntary in order to express a kind of stress and emphasis such as *énnnnorme*. The problem with such kind of words is that they are irregular in the corpus. For example, *énnnnorme* is highly positive but it is not present in the feature set so it is not useful. Highly misspelled reviews tend to be misclassified.

Neutral and mixed reviews. Neutral Reviews such as *le film est visuellement réussi mais le scénario est d'une banalité affligente* are randomly classified. As a matter of fact, reviewers tend to argue their opinion by posting simultaneously positive and negative arguments organized in a concession or a contrast rhetorical form. Lexical classification shows its limits when the abundant polarity of text spans is not coherent to the final retained opinion, typically the case of a reviewer who starts by verbosely listing the film drawbacks and ends by confessing his admiration and concisely posting his favorable judgment.

Ironic expressions and out of scope spans. We noticed that ironic expressions such as *trop fort les gars* that has a negative polarity although it is composed of positive words. Furthermore, since we didn't cope with idiomatic expressions, proverbs and sayings, this issue could in some cases affect the classification. Besides, some reviews contain subjective sentences that describe other satellite subjects that do not concern the reviewer opinion about the movie. For example we could find a description of a particular scene that does not necessarily reflect the global opinion about the movie such as *Monsieur X est très gentil dans le film*. Such out of scope sentences may affect also the classification.

4.1.2 Morpho-Syntactic Features

In further experiments, we appended POS tags to every lemmatized unigram so as to disambiguate same unigrams having different syntactic roles. However, the effect of this information seems barely relevant, as depicted on line (4) of Table 1, the accuracy is only increased by approximately 0.25% up to 92.75%. When applying the negation processing (5) to the same experiment, results were slightly improved (up to 93.00) but still not higher than experiment (3) where no POS tags were used. This entails that ambiguity at the morpho-syntactic level of the reviews does not have much effect on the polarity classification. Thus, we eliminated this feature from our next experiments.

When restricting unigrams features to only adjectives (6), the performance is worse; accuracy is decreased by 6.75% down to 85.75% comparing to (2) and the feature set is reduced by approximately 80%. In order to understand such inconsistency, we look deeper at the accuracy of positive and negative reviews separately. On a one hand, we notice that negative reviews are better classified than positive ones. On the other hand, we have found, in additional experiments, that negative reviews contain relatively an important number of positive adjectives (generally in the negative form). In the first experiment (6) and before processing the negation, these positive adjectives are defined as negative features in the training model, which induces a false decision when classifying positive reviews containing these positive adjectives. However, in the second experiment (7), even after negation processing, the results didn't improve which obviously entails that the scope of processed negation didn't capture the adjectives and was mostly local to the verbal phrase. This last experiment is in contradiction with the results of [HM97] but confirms the results of [PLV02].

As we have already described in section 3.2, negation is sucessfully detected only at the verbal phrase level, but hardly tackled at the other levels such as noun and adjective phrases. Since deep syntactic dependency analysis of reviews is quite a costly task, it is difficult in this case to capture the adjectives related to the negated verb. Heuristic rules discussed previously to resolve this issue such as defining a window of words around the negated verb is not likely to give satisfactory results.

4.1.3 Semantic Features

Apart from the lexical and the POS features, we extend in our experiments the features set to words polarity extracted from SentiWordNet and formulated as a score representing the overall negativity and positivity of words in the reviews. As shown on the table 1 experience (8), results are improved by 1.75% up to 93.25% compared to lemmatized unigrams experiment (2). The main reason of such a barely perceptible improvement is the failure of extracting polarity information of words from SentiWordNet: among 2000 adjectives, we got the polarity information of only 800 entries in SentiWordNet (40% success). This extraction problem is mainly due to the following problems.

Translation errors. We translate words from French to English so as to cope with SentiWordNet interface. However, the quality of translation significantly affects the results of semantic polarity extraction; this is mainly due to the following reasons.

- The bilingual translator doesn't preserve the POS of words. For example, the *noun méchant* is translated into *wicked* which is implicitly an *adjective* and not a *noun*. Since the translator does not reveal information about the POS change after translation, *wicked* is assumed to be a *noun*. However, the *noun wicked* doesn't exist in SentiWordNet.
- Moreover, even if the translation is correct, it happens that the two parallel words do not share the same semantic orientation across both languages due to a difference in common usage, for instance the French positive adjective *féériques* is translated into the negative English adjective *magical*; the French positive adjective *magique* is translated into the negative adjective *magic* as found in SentiWordNet[5].

---

[5] Wei and Pal have come across similar problems when they used an annotated corpus in English for sentiment classification in Chinese [BW10]. They proposed a cross lingual adaption approach to minimize noise introduced by automatic machine translation based on the use of key 'reliable' parts from the translations. They applied in addition a structural correspondence learning to find a low dimensional representation shared by the two languages.

Lemmatization and POS tagging errors. Misspellings are not standard unigrams and hence could not be found in SentiWordNet. Reviews containing a large number of misspellings would have their overall polarity incorrect. In addition, misspellings and other lexical errors (for example punctuation, use of parenthesis *permanente(c' est* and composed words *as-tu-vu*) could significantly affect the results of lemmatization and POS tagging tasks elaborated by TreeTagger. In fact, TreeTagger is not implemented to cope with everyday French language as found in spontaneous movie reviews.

Negation. As shown in the last experiment (9), the integration of negation processing didn't improve the results as expected. Although negation was found to give a relatively significant effect in experiment (3), a slightly decreasing global performance (92.75%) is shown when coupled with polarity feature, which as discussed earlier have shown disappointing performance due to translation imprecision. As a reminder, the negated verbs, nouns and adjectives would have their extracted polarity score from SentiWordNet inverted, for instance the adjective *réussi* holds a negative polarity in the context of *le film n' est pas du tout réussi*. We finally explain such an outcome by two reasons (i) negation didn't properly capture adjectives (considered as the most subjective lexicon) (ii) bilingual translation of subjective lexicon was not very precise.

5 Conclusions

In this paper, an unsupervised approach to sentiment analysis of French movie reviews in a bilingual framework was described. It has been shown that the combination of lexical, morpho-syntactic and semantic features achieves relatively good performance in classifying French movie reviews according to their sentiment polarity (positive, negative). Several problems having an effect upon the results of the classification were highlighted and potential solutions were discussed.

In order to extract the semantic orientation of words from SentiWordNet, we went through a standard word-translation process. Although translation does not necessary preserve the semantic orientation of words due to the variation of language common usage especially when it comes to spontaneous reviews on the web, and in spite of all its side effects, it has been argued that dictionary-based approach could contribute to achieve better results. Even if our first experiments showed little significance, further improvements have been proposed accordingly, particularly concerning negation processing.

In future evaluations, the method will be analyzed within larger training and test sets. Further shallow linguistic analysis will be elaborated such as spelling correction, more elaborated negation processing, WSD and elimination of out of scope text spans from reviews. In addition, the translation task will be improved with the use of French-English EurowordNet so as to apply translation in the sysnset level before querying sentiWordNet database.

References

[AB06]   Andreevskaia, A., Bergler, S.: Mining wordnet for fuzzy sentiment: Sentiment tag extraction from wordnet. In: Proceedings of Conference of the European Chapter of the Association for Computational Linguistics, EACL 2006 (2006)

[ACS08]   Abbasi, A., Chen, H., Salem, A.: Sentiment analysis in multiple languages: Feature selection for opinion classification in web forums. ACM Transactions on Information Systems (TOIS) 26(3), Article 12 (June 2008)

[BW10]   Pal, C., Wei, B.: Cross lingual adaptation: an experiment on sentiment classifications. In: Proceedings of the ACL 2010 Conference Short Papers, pp. 258–262 (2010)

[Den08]   Denecke, K.: Using sentiwordnet for multilingual sentiment analysis. In: Proceedings of the IEEE International Conference on Data Engineering (ICDE 2008), pp. 507–512 (2008)

[ES05]   Esuli, A., Sebastiani, F.: Determining the semantic orientation of terms through gloss classification. In: Proceedings of CIKM 2005, pp. 617–624 (2005)

[ES06]   Esuli, A., Sebastiani, F.: Sentiwordnet: a publicly available lexical resource for opinion mining. In: In Proceedings of the 5th Conference on Language Resources and Evaluation LREC, vol. 6 (2006)

[GA05]   Gamon, M., Aue, A.: Automatic identification of sentiment vocabulary: exploiting low association with known sentiment terms. In: Proceedings of the ACL 2005 Workshop on Feature Engineering for Machine Learning in Natural Language Processing. Association for Computational Linguistics (July 2005)

[Gam04]   Gamon, M.: Sentiment classification on customer feedback data: Noisy data, large feature vectors, and the role of linguistic analysis. In: Proceedings of the 20th International Conference on Computational Linguistics, pp. 611–617 (August 2004)

[HL04]   Hu, M., Liu, B.: Mining and summarizing customer reviews. In: Proceedings of Knowledge Discovery and Data Mining, KDD 2004 (2004)

[HM97]   Hatzivassiloglou, V., McKeown, K.R.: Predicting the semantic orientation of adjectives. In: Proceedings of the 8th conference on European Chapter of the Association for Computational Linguistics, pp. 174–181 (1997)

[HR10]   Hassan, A., Radev, D.: Identifying text polarity using random walks. In: Proceedings of the 48th Annual Meeting of the Association for Computational Linguistics, pp. 395–403 (2010)

[JKdR04]   Mokken, R.J., Kamps, J., Marx, M., de Rijke, M.: Sentiwordnet: a publicly available lexical resource for opinion mining. In: Proceedings of the 5th Conference on Language Resources and Evaluation, LREC 2004, vol. 4, pp. 1115–1118 (2004)

[Joa98]   Joachims, T.: Making large-scale svm learning practical. ACM Transactions on Information Systems, TOIS (1998)

[JP01]   Booth, R.J., Pennebaker, J.W., Francis, M.E.: Linguistic Inquiry and Word Count (LIWC): LIWC 2001. Erlbaum Publisher, Mahwah (2001)

[KH04]   Kim, S.-M., Hovy, E.: Determining the sentiment of opinions. In: Proceedings of the 20th International Conference on Computational Linguistics (COLING 2004), pp. 1367–1373 (August 2004)

[MTO05]   Matsumoto, S., Takamura, H., Okumura, M.: Sentiment classification using word sub-sequences and dependency sub-trees. In: Ho, T.-B., Cheung, D., Liu, H. (eds.) PAKDD 2005. LNCS (LNAI), vol. 3518, pp. 301–311. Springer, Heidelberg (2005)

[NSS07]   Nastase, V., Sokolova, M., Shirabad, J.S.: Do happy words sound happy? a study
           of the relation between form and meaning for english words expressing emotions.
           In: Proceedings of Recent Advances in Natural Language Processing (RANLP
           2007), pp. 406–410 (2007)
[PLV02]   Pang, B., Lee, L., Vaithyanathan, S.: Thumbs up? sentiment classification using
           machine learning techniques. In: Proceedings of the Conference on Empirical
           Methods in Natural Language Processing (EMNLP), pp. 79–86 (July 2002)
[Sch94]   Schmid, H.: Probabilistic part-of-speech tagging using decision trees. In: Pro-
           ceedings of the International Conference on New Methods in Language Process-
           ing, pp. 44–49 (1994)
[SV04]   Strapparava, C., Valitutti, A.: Wordnet-affect: an affective extension of wordnet.
           In: Proceedings of the 4th International Conference on Language Resources and
           Evaluation (LREC 2004), pp. 1083–1086 (May 2004)
[TIO05]   Takamura, H., Inui, T., Okumura, M.: Extracting semantic orientations of words
           using spin model. In: Association of Computational Linguistics ACL 2005, pp.
           133–140 (2005)
[TL02]   Turney, P.D., Littman, M.L.: Unsupervised learning of semantic orientation from
           a hundred-billion-word corpus. Technical Report ERB-1094, National Research
           Council Canada, Institute for Information Technology (2002)
[TL03]   Turney, P.D., Littman, M.L.: Measuring praise and criticism: Inference of se-
           mantic orientation from association. ACM Transactions on Information Systems
           (TOIS), 315–346 (2003)
[TNKS09]   Thet, T.T., Na, J.-C., Khoo, C., Shakthikumar, S.: Sentiment analysis of movie
           reviews on discussion boards using a linguistic approach. In: Proceedings of the
           1st International CIKM Workshop on Topic-Sentiment Analysis for Mass Opin-
           ion Measurement (2009)
[TTC09]   Tang, H., Tan, S., Cheng, X.: A survey on sentiment detection of reviews. Expert
           Systems with Applications: An International Journal 36(7), 10760–10773 (2009)
[Tur02]   Turney, P.D.: Thumbs up or thumbs down? semantic orientation applied to unsu-
           pervised classification of reviews. In: Proceedings of the 40th Annual Meeting of
           the Association for Computational Linguistics (ACL 2002), pp. 417–424 (2002)
[Whi89]   Whissell, C.M.: The dictionary of affect in language. In: Lutchik, R., Kellerman,
           H. (eds.) Emotion: Theory, Research, and Experience, pp. 113–131 (1989)
[WK09]   Wiegand, M., Klakow, D.: The role of knowledge-based features in polarity clas-
           sification at sentence level. In: Proceedings of the Florida Artificial Intelligence
           Research Society Conference (FLAIRS Conference 2009) (2009)
[WWH05]   Wilson, T., Wiebe, J., Hoffmann, P.: Recognizing contextual polarity in phrase-
           level sentiment analysis. In: In Proceedings of the Conference on Empirical
           Methods in Natural Language Processing (EMNLP 2005), pp. 347–354 (October
           2005)
[YH82]   Yu, H., Hatzivassiloglou, V.: The potts model. Reviews of Modern Physics 4(1),
           135–268 (1982)
[YH03]   Yu, H., Hatzivassiloglou, V.: Towards answering opinion questions: Separating
           facts from opinions and identifying the polarity of opinion sentences. In: Pro-
           ceedings of the 2003 Conference on Empirical Methods in Natural Language
           Processing, EMNLP 2003 (2003)
[YNBN03]   Yi, J., Nasukawa, T., Bunescu, R., Niblack, W.: Sentiment analyzer: Extracting
           sentiments about a given topic using natural language processing techniques. In:
           The Third IEEE International Conference on Data Mining (2003)

Query Building in a Distributed Semantic Indexing System

Claude Moulin and Cristian Lai

Abstract. This chapter focuses on the semantic indexing of resources in a Peer to Peer network. Keys used for indexing and the corresponding urls of indexed resources are stored in a distributed hash table scattered on the single peers of a community. A key is a semantic description of resources and can be considered as a small knowledge base in that it refers to concepts and properties belonging to ontologies. In this kind of index a key used for resource retrieval must be identical to a key used for indexing the resource. Therefore, it is necessary to publish a resource with several keys allowing its further retrieval in different contexts. For that purpose, we have defined an expansion mechanism of the keys used for indexing. With different examples, we present the main cases of expansion that define the retrieval context of a resource.

1 Introduction

In the world of communities, people are strongly motivated in sharing resources with respect to their interests. In our research, we mainly deal with communities whose members are scattered in a distributed Peer to Peer (P2P) network where any participant is called Peer or node. In our case, people are not interested in maintaining direct contacts and exchanging messages, or having activities typically facing centralized approaches like in current social Web platforms.

Resources interesting the members of such communities can have different types like documents or notes and can have different formats such as text, image or video. The main activities that community members do regularly, on one hand, concern the indexing of own resources and their publication in the P2P network and, on the

Claude Moulin
University of Technology, Heudiasyc CNRS, Compiègne, France
e-mail: claude.moulin@utc.fr

Cristian Lai
CRS4, Center of Advanced Studies, Research and Development in Sardinia,
Parco Scientifico e Tecnologico, Ed. 1, 09010 Pula (CA), Italy
e-mail: clai@crs4.it

V. Pallotta, A. Soro, and E. Vargiu (Eds.): Advances in DART, SCI 361, pp. 109–128.
springerlink.com   © Springer-Verlag Berlin Heidelberg 2011

other hand, the retrieval of resources from the same network. The community nature is not relevant because the solution we propose is not specific to a certain domain.

The main problem raised by this situation is twofold: (i) what kind of indexing system is more suitable for the members of those communities that are not specialist of software installation and don't want to have a complex system to maintain; (ii) how to retrieve a document thanks to an index which is distributed on different peer computers within a response time as short as possible. In such a Boolean index [20], keys used for resources retrieval must be equals to keys used for publication. The concomitant problem, which is not specific of this approach but is more crucial in these circumstances, is the difference of context between the publication and the retrieval of a resource.

We propose to develop some aspects of the solution we brought to this problem. Our studies address techniques able to create an index of managed distributed resources. The index is distributed among the P2P network via a data structure called Distributed Hash Table (DHT) [4] [23]. The use of a P2P system is justified due to the most relevant features such as decentralization, scalability, security. Any Peer needs to coordinate with only a few other nodes in the system, most commonly $O(\log n)$ of the n participants, so that only a limited amount of work needs to be done for the community activities. The system should function efficiently even with thousands or millions of nodes, assuring a good scalability. Each node of the network contains a portion of the whole data structure. The index is composed of entries that are pairs of data (*key*, *value*). In traditional file sharing systems the *key* is created through the title of the resource (e.g. the title of a song). In our case the *key* is created starting from the semantic description the user gives to the resource. The *value* is the access point to the resource. It is created with the URL of the resource that allows its further access. We have chosen a semantic indexing based on ontologies. The choice of the ontologies is free and these documents are also published in the network by expert members, as the other resources. All the keys used for representing a document in the index represent its semantic description and are written in a language based on RDF. A resource may be indexed by more than one key and a key may index several resources. We can consider the index as a distributed knowledge base in that values of the entries are written in a format based on RDF triples that we have adapted for our purposes. First, the user uses the system to creates the meta-information that describes the content of a document and that includes its personal point of view on the document; then, the system creates the keys of indexing and the user can publish the resource in the network. The system stores in the DHT the pair (key, URL). Keys for resources retrieval are created in the same way. Resources management is essentially publication and retrieval. These activities require a description from the user and the system provides the same keys generation method for both activities. Anyway, it is necessary to take care on the differences between a publication context and a retrieval context. In fact in DHT based systems, keys used for publishing must be equal to those used for retrieval. The solution we propose is to foresee during the publication of a resource, different reasonable retrieval situations, and therefore different queries to which the resource should respond positively. We use a generic reasoning based on the ontologies structures involved in the semantic description of a resource.

In this chapter we explain how to create a key of indexing. The same system of keys is used to submit a query to the system to retrieve a resource or to provide a description for the publication of a resource. We describe the process of resources description that ends with the creation of a key. The description is made upon the elements of domain ontologies provided by the system. No restrictions are imposed so that the user chooses the proper ontologies needed for the description.

2 Related Work

Our system has to deal with technologies combining P2P network, semantic index-ing and domain specific retrieval systems. Progresses in Semantic Web, peer to peer, natural language processing, is leading to new forms of collaboration and social se-mantic information spaces.

AgentSeeker [16] is a multi-agent search engine aiming at managing enterprises knowledge bases. An ontology agent is devoted to manage the enterprise domain in a semantic way. The goal is to make document retrieval a more intelligent process, finding texts which are semantically bind to the user's query. The core behaviour of *AgentSeeker* is to parse text files and to keep a database for storing extracted infor-mation. Every record includes the URIs of the ontologies supported and a measure of the affinity, in term of percentage of words of the document which are also con-tained in the ontology. *AgentSeeker* has only the textual content available, because the author's opinion given by tagging the resources become complicated due to the impossibility of modifying a file or to the difficulty to manually catalogue thousands of documents. In case of specific domains, which *AgentSeeker* considers, it is not prohibitive to think to manually catalogue personal resources if specific tools are provided. It is not necessary to modify the file content in that a semantic description can be associated. In our approach we provide such a tool that associates a semantic description to any kind of documents. As stated in [18], ontologies allow adding semantics to data so that different software components can share information in a homogeneous way. Furthermore, logic can be used in conjunction with such formal representations for reasoning about the information and facts represented as ontolo-gies. We also take the last remarks into consideration in our description expansion system.

In the field of distributed knowledge management, *SA Net* [22] an agent-based system achieves its semantic richness through the use of explicit ontologies to repre-sent resources. *SA Net* further enhances the DHT based resource distribution scheme by using the unique identifier assigned to each ontology as a key to locate the overlay node responsible for maintaining the resource index associated with the underlying ontology. In other words, the ontology-based hashing scheme, utilizes ontologies, instead of resource names, as the hash input to generate the key necessary to distribute the resource among overlay nodes. In our approach we give the same responsibility to all nodes of the network. We have a slightly different meaning of semantic indexing in that we do not directly attach resources to ontologies but create keys whose content refers to ontologies and represents a semantic description of resources. In this way there is a symmetric approach for keys creation in both publishing and retrieving resources in the DHT.

RDFGrowth algorithm [24] introduces a scenario of a peer to peer network where each peer has a local RDF database and is interested in growing its internal knowledge by discovering and importing it from others peers in a group. It is important to consider this kind of approach in order to define a mechanism of queries based on SPARQL formalism. This kind of queries requires RDF knowledge base. The problem should be to distribute a centralized knowledge base on different nodes in order to satisfy a query by accessing only one node. In our scenario the DHT stores the distributed knowledge base. The process starts by browsing several ontologies useful for a user to index or search for resources. The system builds the indexing keys through user inputs. The types of allowed requests determine the types of indexing keys and routing algorithms. In a centralized case a compound query is an investigation on an index organized as a knowledge base (looking for the triples which suit the query in RDF bases). In our case we have to face issues regarding distributed knowledge bases. A direct interrogation of the overall knowledge base is impossible.

Our work aims at demonstrating the benefit of a semantic indexing engine exploited through a set of tools available for a community of users located in different peers. To create a community of users means to tackle the topic of distributed systems. A first requirement is to have systems independent from any central point of aggregation. Among distributed systems, P2P architecture brings most advantages, among them decentralization, scalability, fault tolerance. It is mandatory to have an efficient distributed data structure to efficiently store and retrieve elements from a huge amount of information; the evident efficiency of DHTs relies in the number of messages exchanged to route a query to its destination. The order of magnitude of this number is $O(\log(n))$, where n is the total number of nodes. In this work the low level layer concerning the P2P applications is built on FreePastry[1], the open-source implementation of Pastry [19], whose significance is guaranties by the support provided by the community of users regularly improving and amending; its features allow for adapting the network to the specific needs. We dont need to deduce new metadata from different structured information, but simply to create an index whose content refer to ontologies. However some reasoning aspects have to be taken into account.

A reasoner is a piece of software able to infer logical consequences from a set of axioms or asserted facts. In practice a reasoner makes inferences either about classes constituting an ontology or individual constituting a knowledge base. Ontology classification arranges classes defined by logical expressions into a hierarchy. This reasoning task is normally related to ontology development.

---

[1] http://www.freepastry.org/FreePastry/

Our approach concerns the query answering with respect to ontology based information retrieval. The Semantic Web requires high-performance storage and reasoning infrastructure in order to match the demand of indexing structured data with the use of ontologies. The major challenge toward building such infrastructure is the expressivity of the underlying standards such as RDF(s) and OWL [10]. For the first case of reasoning (concept classification), we may use different available engines related to ontology languages for the Semantic Web like OWL and RDF(s) [9]. Among the most popular there are RacerPro, FaCT++ and Pellet. In particular, Pellet is an OWL DL reasoner based on the tableaux algorithm [1] developed for expressive Description Logics. An interesting feature of Pellet is its strictly relation to ontology analysis and repair. In fact, as explained in [15] OWL has two major dialects, OWL DL and OWL Full, with OWL DL being a subset of OWL Full. All OWL knowledge bases are encoded as RDF/XML graphs. OWL DL imposes a number of restrictions on RDF graphs, some of which are substantial (e.g., the set of class names and individual names be disjoint) and some less so (every item has a type triple). Ensuring that an RDF/XML document meets all the restrictions is a relatively difficult task for authors, and many existing OWL documents are nominally OWL Full, even though their authors intended for them to be OWL DL. Pellet incorporates a number of heuristics to detect DLizable OWL Full documents and repair them, i.e. making them compliant with DL characteristics. In our work we don't take care of the design of ontologies but we assume to find well structured and ready to use ontologies. We are not interested in ontology analysis and repairing. However, manual semantic indexing is only possible with named concepts with clear descriptions. We only propose this kind of concepts in our tools for building semantic descriptions. We cannot deal with anonymous concepts built on logical expressions.

The second case of reasoning based on the structure of the ontology applies the semantics rules of OWL to a knowledge base. This adds new assertions to the knowledge base (the first case adds new axioms to the ontology). In [10] the authors explain two principle strategies for rule-based inference, forward-chaining and backward-chaining. Their approach is based on inferred closure and known as materialization. Through the inferred closure, a knowledge base is extended with all the facts inferred by the application of semantic rules. In our case, the system builds a semantic description of a resource. This description is a small knowledge base that has the form of a tree of nodes. The root element represents the resource to index. Other nodes are ontology identified elements or anonymous local resources allowing to create a path from the root node (as shown in the next figures). We also need an expansion mechanism inferring from only a few semantic rules, mainly relative to subsumption. The application of other semantic rules would infer facts about anonymous elements contained in the small and local knowledge base. It is useless because it is not concerning the resource to be indexed. Our work intends to generate materialization within a virtual knowledge base created each time for a specific case.

3   Semantic Indexing

Indexing is the process of creating or updating an index. Starting from a list of resources this allows to create a correspondence between identifiers and resources. There are many samples of indexes. The index of a book is the association between words or expressions (identifiers) and page numbers (resources). We generally consider two kinds of models for indexing resources: *boolean* [21] and *vectorial*. In the *boolean* model, the index of documents is an inverse file which associates to each keyword of the indexing system the set of contained documents. In the *vectorial* model a document is represented by a vector which dimensions are associated to the keywords occurring in the document and the coordinates correspond to the weights attached to the keywords in the document thanks to a specific calculus. A request is also a vector of the same nature. The system answers a request with the list of documents which present a sufficient similarity with the request thanks to a specific measure based on the vectors coordinates.

In centralized indexing both models are available. However, in the case of P2P networks, the index must be distributed among the peers and the numbers of queries sent to the system when searching for resources should be minimized, because they are time consuming. For respecting this constraint, the model of a distributed index is necessarily Boolean.

Documents may have different types (text, audio, video, archive, etc.) and different storing formats. To share and retrieve the resources it is necessary to give them a proper description. The traditional file sharing diffusion is based on the resource title. We cannot assume that the title could be the right way to identify a resource because in our case resources are created locally and their meaning is not universally known. So it is necessary to find another way for describing the resources.

We can consider a sort of tagging which consists in associating keywords to resources. In many cases keywords bring to an ambiguous description as in the following example: eg. the resource is titled *An interesting overview of Java.* and we use the keyword *Java* as description. What does it mean? Does *Java* refer to the island of Indonesia or does it refer to the programming language developed at Sun Microsystems? This solution is not satisfactory because we cannot accept that a system based on this description type lets too many ambiguous cases. We propose a fine mechanism of description of the resources. We consider semantic descriptions built on elements extracted from ontologies.

The Semantic Web propose standards allowing machines to understand the meaning (or "semantics") of information. It defines a set of well supported languages such as RDF, RDFS, OWL, SPARQL, and related technologies. In our solution, semantic information strictly related to a resource is written in a language based on RDF and is included in its description. We have chosen to use domain ontologies written in RDFS or OWL. We don't take care of the design of ontologies but we assume to find well structured and ready to use ontologies. Manual semantic indexing is possible only if named concepts with clear descriptions are provided.

With respect to the previous example, the choice of the right ontology allows to know if *Java* can be attached to a geographical concept or to a computer science one.

Definition 1. (Semantic Description)
We call Semantic Description the representation of a resource that uses ontological elements.

In our case the building of a Semantic Description is manually performed but a part of the Description could be done automatically with specific software.

Definition 2. (Semantic Indexing)
We call Semantic Indexing the process of associating a Semantic Description to a resource.
The semantic indexing is defined upon the possible queries the system can answer. A query provided by users is a logical expression combining semantic descriptions and Boolean operators (AND, OR). The system answers with the list of documents that satisfy the logical expression. The semantic elementary descriptions included in the user's query must be contained in the index.

3.1 Ontologies

3.1.1 Ontological Elements

Our solution lets open the choice of the ontologies required for the descriptions. However, users of the community should share them else the discovery of the documents published in the network would be impossible. The resource provider is responsible of the choice of the ontology describing the concepts. The manual semantic indexing requires the selection of the ontologies used for building the indexing key. A key may contain several concepts belonging to one or two ontologies. Within an ontology, an element is completely defined by its unique URI[2]. For our system it is enough to insert the URIs of ontological elements for characterizing a document in the key that indexes it.

3.1.2 How to Find Ontologies for Indexing?

The ontologies used for the resource descriptions are published in the P2P network as the other resources. We consider that some expert users have the skills to look for and select ontologies interesting the community members [8]. Our system allows the publication each time a new ontology is useful for the community and can be shared. It also requires a small description and the application domain of this ontology. It is not possible to cancel an ontology. However most of the users are not aware of the existence of ontologies and are not involved in this process. We provide for a

---

[2] http://www.w3.org/Addressing/

tool that helps the users to navigate the ontologies and automatically create the keys corresponding to the publication and retrieval contexts.

We have created a system ontology for implementing some specific cases of resource description (see section 3.4). It also allows to describe an ontology that will serve for indexing resources. A special key, using the system ontology, is created for publication and discovery of the ontologies in the network. In particular the system ontology is also published in the network.

3.1.3 Ontologies Used in the Examples

For describing the different examples presented in this work, we use the following ontologies: an ontology (denoted by *lom*) [6, 7], developed at "Université de Technologie de Compiègne" for representing the domain of learning objects; an ontology (denoted by *p2p-lt*) for describing the concepts of the theory of languages; the system ontology is denoted by *system*.

3.2 Types of Queries

3.2.1 Classical Approach

In ontology engineering it is necessary to answer questions related to modeling practice. Among them we find: *Who does what, when and where?*, *What are the parts of what?*, *What is an object made of?* [5].

In query engineering, studies have to be managed in order to define the types of queries a user can produce. Common queries may have a general purpose like: *What are the documents about the subject Medieval Italy?*. Other queries seems general but refer to a specific domain, theory of languages, such as: *What are the documents written by Chomsky?*, or *What are the documents published by a specific author in 2008?*.

Two queries may be considered dependent. For example in the field of programming languages it is usual to refer to concepts concerning data structures like *stack, list or tree*. Queries like *What are the resources that concern stack?* and *What are the resources that concern data structures?* are not independent when a relation between the concepts of stack and data structure exist in the ontology. In some cases a resource can be described in different ways, so a resource can be associated to more queries.

Even if we consider semantic descriptions of resources, there are some different ways a user can address a query to a system. Moreover the system user interface must be designed in line with the query model accepted by the system. We assume a system accepting domain ontologies eventually augmented with knowledge bases containing the definition of the main individuals of a domain. In our case we consider different ways the user can input a query in a system.

• The user produces a query written in natural language: e.g. *What are the resources that concern stack?*

- The user produces a query containing a sequence of keywords and expressions combined with boolean operators: e.g.
  resources AND *stack*; *resources* NOT *stack*.

This two cases require a text analysis of the query in order to identify the concepts and relations of an ontology that can be associated to the content of the query [12].

The necessary matching between a query content and an ontology content is based on text analysis techniques such as supplied by Lucene [13]. Sometimes a query analyser is able to detect incomplete formulation and may begin a dialogue [17] with the user for adding the missing elements.

These two approaches have some limitations and are not suitable for our aims. In both two last cases there could be noise due to users input which could produce typing errors. Moreover, use of synonymous imply to use more sophisticated text analysis. A system with such technologies is heavy.

We propose a third approach based on description patterns. The user does not formulate herself the query but is guided by a system for building it. The system creates in the background the real query in a specific internal language and sends it.

3.2.2   Our Approach

The user must first identify the type of resources she is looking for. It represents the category of resources she has in mind. This is made choosing the proper ontology related to the domain of the resource. It is the ontology that contains at least one concept which denotes the resources the user is interested in. Figure 1 shows the Query building process. The approach considers resources as instances of well identified concepts (the possible types of the resources). The system proposes this list of concepts. Once the initial concept is chosen, a list of properties related to the type of the resource is proposed to the user. The process goes on following step by step a path in an ontology and stops when the end of the path is reached. The system creates a description resulting of the steps followed. A query provided by users is a logical expression combining semantic descriptions and Boolean operators (AND, OR). For instance the user thinks to the following query: *What are the documents written by Chomsky?*

The system guides the user to navigate the chosen ontology, recognizes the minimal path of the ontology that represents the meaning of the user query and builds the real query. In the example the real query can be translated in natural language by *What are the documents having an author whose name is Chomsky?*

Figure 2 shows the path followed by the user and complementary information found in an ontology (or in a knowledge base).

The path is a graph where the nodes are individuals or concepts and the arcs are properties of the ontology (or knowledge base).

3.2.3   Fundamentals

We consider two fundamental types of queries. The first one (called *Resource Query Type* in the rest of the document) aims at finding resources from elements of

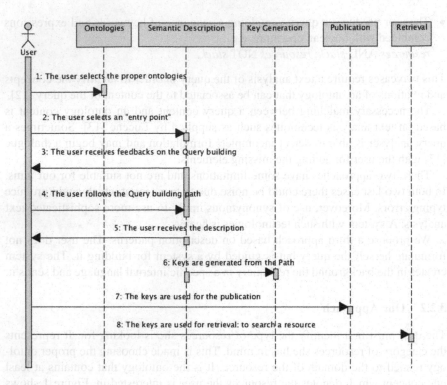

Fig. 1 The Query Building brings to create a key both for publication and retrieval

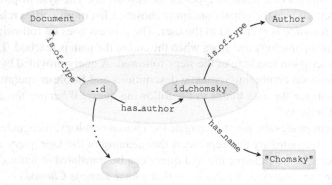

Fig. 2 Query building

description that concern the resources themselves and not directly their content. For instance: *What are the resources written by Chomsky? I search difficult exercises.* We consider that the resources concerned by the queries of this type can be represented by instances of a specific concept of a domain ontology: *Theory of Languages* for the first example and *Learning Object* for the second one. We call *Resource Type* the initial concept that the user must identify for denoting the resources she looks for. The properties useful for interpreting the user queries have to be chosen in the ontology of the resource type. This kind of queries relies on only one ontology.

The second type (called *Content Query Type* in the rest of the document) aims at finding resources from elements of description that concern their content. For instances: *What are the resources about Chomsky? What are the resources about grammar?* Both queries concern the topic of the content of the resource. The queries ask for a resource *about* something and not *written by* somebody. It is not possible to find a unique ontology for describing at the same time the essence of a resource and its content. For that purpose we needed and we built a specific ontology (the *System Ontology*). Instances of its Document concept (*system:Document*) represent the resources concerned by the queries of this type. The system:Document is the domain of some properties useful for accessing the elements of other ontologies that give a description of the resource content. Elements of a domain ontology for the last two examples *Theory of Languages* can represent the content of the resources concerned by the queries of this type. This kind of queries relies on two ontologies.

3.2.4   Resource Query Type

The first type of queries aim at finding resources from elements of description concerning the resource themselves and not their content. Let's consider the query: *What are the documents written by Chomsky?* that a user may imagine.

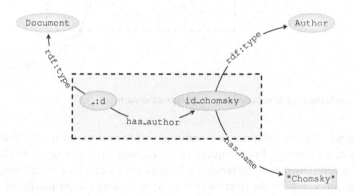

Fig. 3 Knowledge base associated to a query about resources

Our system helps her to discover the concept that the resources are instances of. In this case the user chooses the concept of Document (among the ontology entry points proposed by the system) occurring in the ontology about *Theory of Languages*. The ontology contains a property which domain is the Document concept and that denotes the meaning of the query: *has_for_author* translating *written by* leads the Author concept.

The process goes on following the next steps in the ontology. The augmented ontology contains some individuals of the Author concept. The system presents these individuals and the user can chose *id_chomsky*, verifying with the displayed attributes that it is the right author. The process is finished and the end of the path is reached. In this case the choice of a property and an individual has determined all the elements contained in the user query.

Figure 3 shows the ontological elements related to the query. The rectangle delimits the path that the system uses to build the key that is needed to look for the resources.

3.2.5   Content Query Type

This type of queries aims at finding resources from elements of description that concern their content.

For instance the user imagine the query: *What are the documents about the author Chomsky?*. She is looking for resources whose content gives information about the author Chomsky.

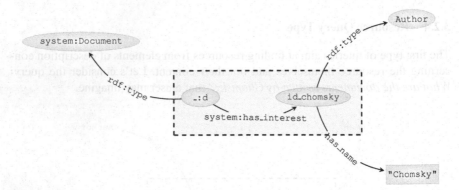

Fig. 4 Knowledge base associated to a query about content

This kind of queries require two ontologies. The system shows the entry points of the System Ontology. For the query of the example, the user needs to choose the concept system:Document. Instances of the concept system:Document represent the resources concerned by the queries of this type. Among the properties of the System Ontology the user selects *system:has_interest* because its domain contains the concept *system:Document*. For representing the content of the resources concerned by the query a second ontology is necessary: the domain ontology *Theory of Languages*. The system shows the possible elements of the second ontology that can be contained in the range of the property *system:has_interest*. The user selects the individual of the concept *Author* identified as *id_chomsky*. This individual has for name "Chomsky". The process is finished as the end of the path is reached. The elements contained in the user query have been determined by the choice of a property and an individual.

Figure 4 shows the ontological elements related to the query. The rectangle delimits the path that the system uses to build its internal query.

Now, let's consider the query: *What are the documents about Grammar?*. The user is looking for resources whose contents give information about the topic Grammar.

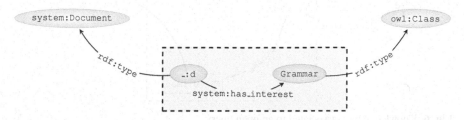

Fig. 5 Knowledge base associated to documents about Grammar

The first step of the process is the same as for the last example: the user selects an individual of the *system:Document* concept and the *system:has_interest* property. In the second ontology, *Theory of Languages*, the user finds and select the concept *Grammar*. The end of the path is reached, then the process is finished. The elements contained in the user query have been determined by the choice of a property and a concept.

Figure 5 shows the ontological elements related to the query. For building its internalquery the system uses the path delimited by the rectangle.

3.2.6 Open and Closed Queries

Our approach allows to build *closed* queries i.e. all their elements are available in the ontology (or in the knowledge base).

However we have considered a particular case of an *open* query where the user can add the value of a property.

For instance, if the user intends to ask: *What are the documents written by XXX?*, where the *XXX* value is not in the knowledge base, she is allowed to add it. Figure 6 shows that the user has added "D.Ullman" as an author name.

With open queries we give the user the possibility to input a keyword manually. It's important to observe that such keyword is not out of any context but is provided as a value of a property of an ontology which is necessarily the last step of the query building process.

3.2.7 Other Queries

Other queries are mere complex because they concern a path involving more nodes (see figure 7).

Let's consider the following query: *What are the documents very difficult?* The query aims at finding resources giving elements of description concerning the

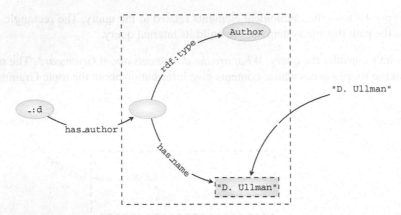

Fig. 6 Knowledge base associated to an open query

Fig. 7 Other queries involve more nodes

essence of the resources themselves. This is the case of "Type one" queries. The meaning of the user query refers to those resources that are documents; the documents are identified as those with an attribute of difficulty; the level of difficulty is "very difficult". The ontological elements required for interpreting the user query rely on the domain ontology *Learning Object Model, LOM*. Among the entry points the system proposes, the user selects *Learning Object*. It is the concept which individuals can represent the resources searched by the user with the query of the sample. For express the meaning of difficulty the LOM ontology provides the concept *Difficulty*. It is not possible a direct connection with a property between an individual of Learning Object and an individual of Difficulty. In fact there is no property within the LOM ontology which domain is the Learning Object concept and range is the Difficulty concept. It is necessary a path with more steps.

After choosing the entry point *Learning Object*, the system proposes the properties with domain the Learning Object concept. The user selects *has_lomEducational*. This property has as range the concept LomEducationalCategory. In the next step the user selects the property *has_difficulty* and an individual of the class Difficulty as range of this property. This individual has for label "Very difficult". The process is finished and the end of the path is reached.

The ontological elements related to the query are shown in figure 9.

3.2.8 Query Extension

A query is an extension of another query when the resources it asks for are concerning a more general category of elements. The extension of a query is obtained from the elements of the first query and a reasoning about the concerned ontologies. The extended query relies on another path created during the search of relations in the ontologies about the elements of the first path.

If the query is: *What are the documents about Stack?*

An extended query is: *What are the documents about Data Structure?*

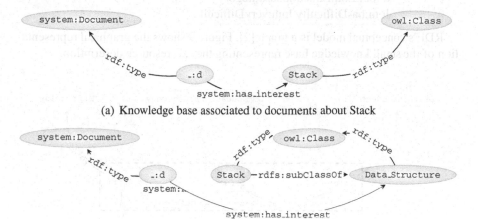

(a) Knowledge base associated to documents about Stack

(b) Knowledge base associated to documents about Data Structure

Fig. 8 Extension of a Query

The documents concerning *stack* also concern *data structure* when the concept Data_Structure subsumes the concept Stack.

3.3 Keys for Publication and Retrieval Contexts

Let's consider an example related to the ontology denoted by *lom*, concerning the learning object domain. We can say that a resource has for *difficulty* level, *very difficult*.

Using the N3 notation [3] we synthesize the resource description as follow:

```
[ a lom:LearningObject ] lom:hasLomEducational
[a lom:LomEducationalCategory ;
    lom:hasDifficulty lom:veryDifficult .]
```

We are speaking about a resource whose *lom:LomEducationaCategory* has for *lom:difficulty lom:veryDifficult*. We create a virtual knowledge base with two instances _*:lo* and _*:lec*. _*:lo* is an instance of the class *lom:LearningObject*, subject of the relation *lom:hasLomEducational* with the object _*:lec*, that should be specified.

```
_:lo
    rdf:type lom:LearningObject ;
        lom:hasLomEducational _:lec .
```

_:lec gives the level of difficulty and is subject of the relation *lom:LomEducationalCategory* with *lom:veryDifficult*, an instance of the class *lom:Difficulty*.

> _:lec
>   a lom:LomEducationalCategory ;
>   lom:hasDifficulty lom:veryDifficult .

RDF's conceptual model is a graph [2]. Figure 9 shows the graphical representation of the small knowledge base representing the _:lo resource description.

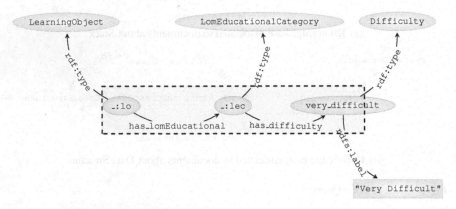

Fig. 9 The RDF graph of Learning Object

We define the publication context of the _:lo resource with the following triples:

> _:lo
>   rdf:type lom:LearningObject ;
>   lom:hasLomEducational _:lec .
> _:lec
>   a lom:LomEducationalCategory ;
>   lom:hasDifficulty lom:veryDifficult .

The resource is identified by its type and the composition of properties where it is involved. A description corresponds to a list of RDF triples containing a path, i.e. a sequence where the object of a triple is the subject of the following one. The other triples denote the link between an instance and a concept. The publication context is used to create the key associated to the resource in the distributed index. The anonymous nodes identifiers are cancelled.

> Key_1:
>   {rdf:type,lom:LearningObject,}
>   {lom:hasLomEducational,}
>   {lom:hasDifficulty,lom:veryDifficult}

For retrieving a resource, a key must be supplied and must be equal to the key created from the publication context. A retrieval context is the description of a required resource and must correspond to a publication context; the identifiers of the anonymous nodes do not matter. In the example, Key_1 should be created. However, it is interesting that the resource (denoted by _:lo) can be found from other queries. For example, a user may be interested by learning object where the difficulty level has been defined, no matter which level. In this case, the retrieval context is a graph pattern containing a variable for representing the undefined value.

```
_:lo
      rdf:type lom:LearningObject ;
lom:hasLomEducational _:lec .
_:lec
      lom:hasDifficulty ?x .
```

In this case the retrieval key is:

```
Key_2:
      {rdf:type,lom:LearningObject,}
      {lom:hasLomEducational,lom:hasDifficulty}
```

In the same way, we have identified different situations where we define several retrieval contexts from the same publication context. This definition requires an expansion mechanism which is presented in section 3.2.8. Generally, semantic indexing is the attachment of a resource to a concept. In this case, the provider means that the subject of the resource is about a concept defined in an ontology but without any other specification. We have considered this eventuality and created a system ontology with the concept of document and the relation of interest for representing it.

In the following example, the resource _:d is about the concept of Stack (extracted from the ontology on the theory of languages denoted by p2p-lt). For maintaining a DL ontology, we cannot associate a resource to a concept. We consider that the resource concerns any representative instance of the concept.

```
_:d
      a system:Document ;
      system:hasInterest _:s .
_:s rdf:type p2p-lt:Stack .
```

The concept of Stack has a super-concept: DataStructure. Figure 8(b) shows the graphical representation of the small knowledge base representing the _:d resource description.

In order to extend the publication context of the _:d resource, we have to consider the subsumption on the Stack concept. _:s has type Stack and then has also the DataStructure type. We can identify two keys:

```
Key_1:
      {system:hasInterest,rdf:type,p2p-lt:Stack}
Key_2:
      {system:hasInterest,rdf:type,p2p-lt:Data_Structure}
```

Then we reduce them to:

> Key_1:
>   {system:hasInterest,p2p-lt:Stack}
> Key_2:
>   {system:hasInterest,p2p-lt:Data_Structure}

Each resource may be considered as an instance of a concept of one or more ontologies. It is possible to combine several elementary descriptions of resources and to create as many keys.

3.4 The System Ontology

The second type of queries aims at finding resources from elements of description that concern their content. It concerns the topics of the content of the resource. Using this type of query the resource provider means that the subject of the resource is only about a concept defined in an ontology without any other specification. Users must discover the resource when they tell to the system that they are interested in resources concerning this concept. For implementing this type of queries we have created a *System Ontology*. It contains the concept of *Document* and the relation *has_interest* for representing this case.

Using this ontology (denoted by *system*) and the ontology of Theory of Languages domain (denoted by *p2p-lt*), it is possible to describe a document concerning the concept of Automaton with:

> a system:Document ;
> system:hasInterest p2p-lt:Automaton.

The logical level of this representation is OWL Full. Inference in OWL Full is clearly undecidable as OWL Full does not include restrictions on the use of transitive properties which are required in order to maintain decidability. This is not a problem of reasoning because we apply a DL reasoning, avoiding the use of inference engines. Even if it is not our case, we could introduce an anonymous instance of a concept, for instance the p2p-lt:Automaton, for maintaining the DL level. The correct meaning would be: a resource concerning any example of automa.

The system ontology is also useful for characterizing the ontologies used for indexing and that have to be published in the network. The concept *system:Ontology*, subconcept of *system:Document* is created for that purpose. An ontology used for indexing is published in the network under the description: [a system:Ontology].

In case where no ontology is available, the system ontology is also useful for associating some keywords with a resource.

4 Conclusion

In this chapter we have presented an approach for indexing resources in a peer to peer network, where users are interested in their sharing. The resources are in every case semantically indexed via domain specific ontologies downloaded from the network. The semantic information strictly related to a resource represents a point of view of the user on the resource. The semantic index, which can be considered as a distributed RDF knowledge base is inserted in a Distributed Hash Table whose structure guaranties an efficient management of the resources. We consider that the ontologies required by the indexing can be provided by some expert users.

It is necessary to navigate the suitable ontologies in order to understand their different concepts and relations. This operation is time consuming but, nevertheless, it is the price to pay when using a semantic indexing and for profiting of its advantages regarding a keyword indexing. Ontologies allow some reasoning and we have shown how the structure of the index requires taking into account this characteristic for improving the queries. The proposed solution consists in foreseeing during the publication of a resource, different reasonable retrieval situations, and therefore different queries to which the resource should respond positively.

References

1. Baader, F., Sattler, U.: Tableau Algorithms for Description Logics. Studia Logica: An International Journal for Symbolic Logic (2001)
2. Beckett, D, Berners-Lee, T: Turtle - Terse RDF Triple Language. W3C Team Submission (2008), http://www.w3.org/TeamSubmission/turtle/
3. Berners-Lee, T.: Notation 3: A readable language for data on the web (2006), http://www.w3.org/DesignIssues/Notation3.html
4. Chawathe, Y., Ramabhadran, S., Ratnasamy, S., LaMarca, A., Shenker, S., Hellerstein, J.: A case study in building layered dht applications. In: SIGCOMM 2005: Proceedings of the 2005 Conference on Applications, Technologies, Architectures, and Protocols for Computer Communications, vol. 35, pp. 97–108. ACM, New York (2005)
5. Gangemi, A., Presutti, V.: Ontology Design Patterns. In: Handbook of Ontologies, 2nd edn., Springer, Berlin
6. Ghebghoub, O., Abel, M.-H., Moulin, C.: Learning Object Indexing Tool Based on a LOM Ontology. In: Eighth IEEE International Conference on Advanced Learning Technologies, ICALT 2008, Santander, Spain (2008)
7. Ghebghoub, O., Abel, M.-H., Moulin, C., Leblanc, A.: A LOM ontology put into practice. In: Second International Conference on Web and Information Technologies, ICWIT 2009, Kerkennah Island Sfax, Tunisia (2009)
8. Gruber, T.R.: Towards Principles for the Design of Ontologies Used for Knowledge Sharing. In: Guarino, N., Poli, R. (eds.) Formal Ontology in Conceptual Analysis and Knowledge Representation. Kluwer Academic Publishers, Deventer (1993)
9. Mei, J., Parsia, B.: Reasoning Paradigms for OWL Ontologies. Department of Information Science, Freie Universitat Berlin, Techreport (2004)
10. Kiryakov, A., Ognyanov, D., Manov, D.: OWLIM - A Pragmatic Semantic Repository for OWL. In: WISE Workshops, pp. 182–192 (2005)
11. Moulin, C., Barthès, J.-P., Bettahar, F., Sbodio, M.: Representation of Semantics in an E-Government Platform. In: 6th Eastern European eGovernment Days, Prague, Czech Republic (2008)
12. Moulin, C., Bettahar, F., Sbodio, M., Barthès, J.-P., Korda, N.: Adding Support to User Interaction in Egovernment Environment. In: 4th Atlantic Web Intelligence Conference, AWIC 2006, Beer-Sheva, Israel, pp. 151–160 (2006)

13. Apache Lucene Project. Available on Internet at, `http://lucene.apache.org/`
14. Pan, Z.: Benchmarking DL Reasoners Using Realistic Ontologies. In: Proceedings of the OWLED 2005 Workshop on OWL: Experiences and Directions (2005)
15. Parsia, B., Sirin, E.: Pellet: An OWL DL Reasoner. In: 3rd International Semantic Web Conference, ISWC 2004 (2004)
16. Passadore, A., Grosso, A., Boccalatte, A.: An agent-based semantic search engine for scalable enterprise applications. In: Proceedings of the 3rd International Workshop on Ontology, Conceptualization and Epistemology for Information Systems, Software Engineering and Service Science (ONTOSE 2009), vol. 460, pp. 82–94 (2009)
17. Ramos, M., Tacla, C.A., Sato, G., Paraiso, E., Barthès, J.-P.: Dialog Construction in a Collaborative Project Management Environment. In: 14th International Conference on Computer Supported Cooperative Work in Design, CSCWD 2010 (2010)
18. Rodriguez, D., Sicilia, M.-A.: Defining spem 2 process constraints with se-mantic rules using swrl. In: Proceedings of the Third International Workshop on Ontology, Conceptualization and Epistemology for Information Systems, Soft- ware Engineering and Service Science (ONTOSE 2009), vol. 460, pp. 95–104 (2009)
19. Rowstron, A., Druschel, P.: Pastry: Scalable, decentralized object location, and routing for large-scale peer-to-peer systems. In: Liu, H. (ed.) Middleware 2001. LNCS, vol. 2218, pp. 329–350. Springer, Heidelberg (2001)
20. Salton, G., Fox Edward, A., Harry, W.: Extended Boolean information retrieval. Technical Report, Cornell University (1982)
21. Salton, G., Fox, E.A., Wu, H.: Extended boolean information retrieval. Commun. ACM 26(11), 1022–1036 (1983)
22. Sangpachatanaruk, C., Znati, T.: Semantic driven hashing (sdh): An ontology-based search scheme for the semantic aware network (sa net). In: P2P 2004: Proceedings of the Fourth International Conference on Peer-to-Peer Computing, pp. 270–271. IEEE Computer Society, Washington, DC (2004)
23. Stoica, I., Morris, R., Karger, D., Kaashoek, M.F., Bal-akrishnan, A.: Chord- A scalable peer-to-peer lookup service for internet applications. In: SIGCOMM 2001: Proceedings of the 2001 Conference on Applications, Technologies, Architectures, and Protocols for Computer Communications, pp. 149–160. ACM, New York (2001)
24. Tummarello, G., Morbidoni, C., Petersson, J., Puliti, P., Piazza, F.: Rdfgrowth, a p2p annotation exchange algorithm for scalable semantic web applications. In: P2PKM (2004)

Building Distributed and Pervasive Information Management Systems with HDS

Federico Bergenti and Agostino Poggi

Abstract. This paper focuses on building distributed and pervasive information management systems using the novel tool HDS (Heterogeneous Distributed System). First, we frame the presented research in the scope of information agent technology by reviewing relevant work on the subject and by stating the notable role that communication plays in such a field. Then, we introduce HDS and we provide details on its functionality and internals. Finally, we present the RAIS (Remote Assistant for Information Sharing) pervasive information management system and we show how it has been recently retargeted to use HDS as its main communication and distribution vehicle.

1 Introduction

The role of agent technology in the field of information management systems is remarkable mainly because *(i)* it provides solid abstractions and models that help the study and implementation of such systems; and *(ii)* it supports their concrete implementation with mature tools that targets large-scale, decentralized and pervasive systems. The important role of agent technology for this kind of systems is also witnessed by the recent introduction of the idea of *intelligent information agent* to support the implementation of decentralized and scalable information management systems.

A closer look to the relationship between agent technology and information management systems reveals that the agent paradigm has been intensively applied in the development of such systems for taking advantage of various features of agents. For example, we see agents used in information management systems as a

Federico Bergenti
Dipartimento di Matematica, Università degli Studi di Parma
Viale U. P. Usberti 53/A, 43100 Parma, Italy
e-mail: federico.bergenti@unipr.it

Agostino Poggi
Dipartimento di Ingegneria dell'Informazione, Università degli Studi di Parma
Viale U. P. Usberti 181/A, 43100 Parma, Italy
e-mail: agostino.poggi@unipr.it

V. Pallotta, A. Soro, and E. Vargiu (Eds.): Advances in DART, SCI 361, pp. 129–142.
springerlink.com   © Springer-Verlag Berlin Heidelberg 2011

means for supporting communication, coordination and even presentation of information to users (see, e.g. [7]).

In Section 2 we review relevant work on the use of agent technology for information management systems and we stress the role of agents in communication and in support of decentralized, scalable distribution. Agents are ideal tools for implementing software systems with advanced features in terms of decentralized, scalable distribution and therefore they are a key ingredient in next generation information management systems where the amount of managed information is going to overwhelm any centralized approach.

In this scenario of agent-based information management systems capable of effectively coping with unforeseen information overload, we present *HDS (Heterogeneous Distributed System)*, a software framework that simplifies the implementation of decentralized applications by merging the client-server and the peer-to-peer paradigms and we discuss its uses for the implementation of flexible and pervasive information management systems. Section 3 introduces HDS and it provides an overview of its features and internals.

In order to exemplify the role of HDS in the implementation of decentralized information management systems, Section 5 describes *RAIS (Remote Assistant for Information Sharing)*, a peer-to-peer multi-agent system supporting the sharing of information among a community of users connected through the Web. RAIS offers a search facility similar to Web search engines, but it avoids the burden of publishing the information on the Web while providing warranties of controlled and dynamic access to the information. The use of agents in such a system is very important because it simplifies the implementation of three core RAIS services: information filtering, information publishing and management of networks of reputation.

Finally, the paper concludes with Section 6 that summarizes the results and sketches some future research directions.

2 Intelligent Information Agents

Information agent technology [5] is a keyword that arose in the research on software agents around ten years ago to provide concrete responses to the challenges that the Internet scale posed to information management systems. While information agent technology does not bring any new feature to information management systems, it makes them capable of coping with the unforeseen amount of information sources that the open and dynamic nature of the Internet brings, and with the issues that it inevitably implies. This is the reason why information agent technology is inherently interdisciplinary and why it needs to fuse a number of approaches, methods and tools that originated in different research community with diverse aims into a coherent view, e.g., advanced database and knowledge-base systems, distributed information systems, and human-computer interaction (see, for example, [8]).

The main objective of information agent technology is to supply methods and tools for the effective implementation of *intelligent information agents*, i.e., software entities capable of providing value-added interfaces to multiple, heterogeneous information sources that are possibly spread across a network. The type, size and topology of the network are irrelevant and intelligent information agents promote a scalable and decentralized approach that copes with networks of information sources ranging from local Intranets to the whole Internet.

Together with specific features discussed below, intelligent information agents inherit some characterizing features of software agents. The very basic feature that they take over from software agents is that they are not simply a concrete development technology; rather they entail a set of abstractions and metaphors that are not necessarily related to how they are concretely built. This is the reason why we can talk about intelligent software agents while modelling an information management system at different stages of the development cycle and with different levels of detail. We say that the main features that characterize intelligent information agents are the skills that they have and their functional and non-functional properties, rather than the agent-oriented technique used in their implementation.

However, skilled intelligent information agents exhibit their autonomous behaviour by means of a synergic combination of: *(i)* pro-activity, i.e., they are goal-directed rather than statically programmed and they can anticipate future information needs while bringing about their goals; *(ii)* timely reactivity, i.e., they keep information sources under surveillance and they timely react to relevant changes; and *(iii)* social behaviour, i.e., they share goals among groups and they compete and/or cooperate in order to fulfil them.

The main tasks that intelligent information agents perform are intended to anticipate future queries from users and/or other agents and to maintain personalized views of information sources in order to facilitate information fruition. Such tasks require equipping intelligent information agents with notable skills, e.g., retrieve, analyze and fuse information from heterogeneous sources into coherent views that anticipate future information needs. Moreover, most advanced intelligent information agents are also in charge of providing personalized user experiences and they have advanced skills in presenting their views of information to users and other agents.

Many systems of intelligent information agents have been built and deployed both in academic and in industrial settings and intelligent information agents are nowadays best practices for large-scale, decentralized information management systems in industrial control, Internet search, personal assistance, network management, games, and many others application areas (see, e.g., [8]).

The literature proposes a classification of such a wide variety of systems [6] on the basis of the actual skills and on the characteristics of agents as follows:

1. Cooperative: if agents share goals and can cooperatively bring about such goals, possibly exploiting cooperation as a means of optimization;
2. Adaptive: if agents' behaviour can change according to changes in the work environment, which comprises information sources, and/or other agents;
3. Rational: if agents' behaviour is designed to explicitly maximize a measurable utility, which can possibly depend on the behaviour of other agents; and

4. Mobile: if agents travel in the network to get close to information sources in order to, e.g., minimize the amount of data transmitted over the communication network or balance the workload across the network.

The sketched classification supports the identification of key skills that characterize intelligent information agents, as shown in Figure 1. Not all agents exhibit all such skills and not all skills are needed to turn an agent into an intelligent information agent, but such a schematic representation shows the main topics we need to get in touch when dealing with advanced intelligent information agents.

Among the skills shown in Figure 1, the agent research community have been working hardly on the communication skills because *(i)* they are the very basic vehicle for decentralized cooperation, and *(ii)* because they are needed to support open, decentralized multi-agent systems as some standardization body envisaged [5]. The research on communication skills provided solid results, e.g., JADE [2,3], that are nowadays considered consolidated baselines for comparisons. Next section presents a step ahead in this area of research, i.e., a novel tool to support distribution and communication in a multi-agent system that capitalizes and extends the decennial research on JADE.

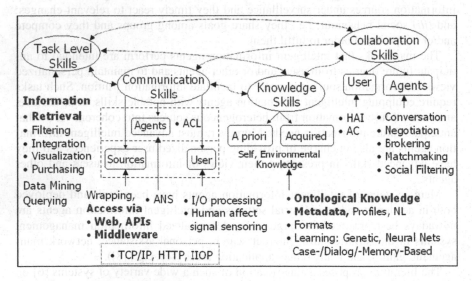

Fig. 1 Basic skills of an information agent (from [7]).

3 HDS

HDS (Heterogeneous Distributed System) is a software framework that aims at simplifying the implementation of distributed applications by merging the client-server and the peer-to-peer paradigms and by implementing all the interactions among all the processes of a system through the exchange of messages.

This software framework allows the implementation of systems based on two types of processes: *actors* and *servers*. Actors have their own thread of execution and perform tasks interacting, if necessary, with other processes through synchronous and asynchronous messages. Servers perform tasks upon request from other processes by composing (if necessary) the services offered by other processes through synchronous messages. While both servers and actors may directly take advantage of the services provided by other kinds of application, only the servers provides services to external applications by simply offering public interfaces (one or even more).

Actors and servers can be distributed on a heterogeneous network of computational nodes, the so called *runtime nodes*, for the implementation of different kinds of application (see Figure 2). In particular, actors and servers are grouped into runtime nodes that make up a platform. An application can be obtained by combining pre-existent applications by means of a federation.

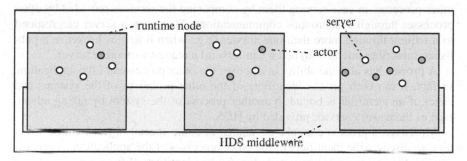

Fig. 2 The architecture of an HDS distributed system.

3.1 Software Architecture Model

The software architecture of an HDS application can be described through the following four different models: *(i)* the *concurrency model*, that describes how the processes of a runtime node can interact and share resource; *(ii)* the *runtime model*, that describes the services available for managing the processes of an application; *(iii)* the *distribution model*, that describes how the processes of different runtime nodes can communicate; and *(iv)* the *communication model* that provides a simple model for defining the data structures that are used as contents of messages.

The concurrency model is used for both defining the processes that build an application and for supporting their interaction. This model is based on four main elements: *process, selector, message* and *filter*.

A process is a computational unit able to perform one or more tasks taking advantage, if necessary, of the tasks provided by other processes. To facilitate the cooperation among processes, a process advertises itself by making its description available to the other processes. The default information contained in a description is represented by the process identifier and the process type; however, a process may introduce some additional information in its description.

A process can be either an actor or a server. An actor is a process that can have a proactive behaviour and so it can start the execution of tasks without any request from other processes. A server is a process that is only able to perform tasks in response of the request of other processes.

A process interacts with the other processes through the exchange of messages based on one of the following three types of communication: *(i)* synchronous communication, the process sends a message to another process and waits for its answer; *(ii)* asynchronous communication, the process sends a message to another process, performs some action and then waits for its answer; and *(iii)* one-way communication, the process sends a message to another process, but it does not wait for an answer.

In particular, while an actor can start all the three previous types of communication with all the other processes, a server can only respond to the requests of other processes in case serving them by composing the services provided by other processes through synchronous communications. Moreover, a server can respond to a request through more than one answer (e.g., when it acts as broker in a publisher-subscriber interaction) and it can forward a request to another server.

A process has also the ability to discover the other processes of the application. In fact, it can both get the identifiers of the other processes of the systems and check if an identifier is bound to another process of the system by taking advantage of the registry service provided by HDS.

Moreover, a process can take advantage of some special objects, called *selectors*, to discover the identifiers of the other processes of the application. A process can ask the runtime registry service for: *(i)* the identifiers of the runtime nodes of the application; *(ii)* the identifiers of the processes running on a runtime node of the application; and *(iii)* the identifiers of the subset of the processes of a runtime node that provide some specific feature. In fact, a selector allows the definition of some constraints on the features of the processes of the application, e.g., the process must be of a specific type or the process must be located in a specific runtime node, and the runtime registry service is able to apply such constraints on the registered processes. Selectors are also used by the processes for extracting from their input message queue the message they need for completing the current task. In this case, the selector defines some constraints on the features that are necessary for identifying the waited message, e.g., the reply to a message sent by the process or a message with a specific kind of content.

A message contains the typical information used for exchanging data on the network, i.e., some fields representing the header information, and a special object, called content, that contains the data to be exchanged. In particular, the content object is used for defining the semantics of messages, e.g., if the content is an instance of the `Ping` class, then the message represents a ping request, and if the content is an instance of the `Result` class, then the message contains the result of a previous request.

Normally, a process can communicate with all the other processes and the act of sending messages does not involve any operation that is not related to the delivery of messages to the destination; however, the presence of one or more filters can modify the normal delivery of messages.

A filter is a *composition filter* [4] whose primary aim is to define the constraints on the reception/sending of messages; however, it can also be used for manipulating messages (e.g., their encryption and decryption) and for the implementation of replication and logging services.

Each process has two lists of filters: the ones of the first list, called *input filters*, are applied to the input messages and the others, called *output filters*, are applied to the output messages. Figure 3 shows the flow of the messages from the input filters to the output filters. When a new message arrives or must be sent, the filters of the appropriate list are applied to it in sequence; such a message is stored in the input queue or it is sent only if all the filters succeed.

The runtime model defines the basic services provided by the middleware to the processes of an application. This model is based on four main elements: *registry*, *processer*, *dispatcher* and *filterer*.

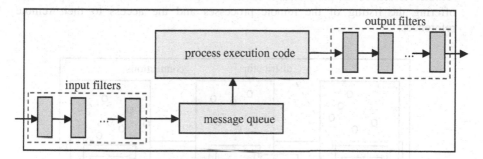

Fig. 3 Flow of the messages inside an HDS process.

A registry is a runtime service that allows the discovery of the processes of the application. In fact, a registry provides the binding and unbinding of processes with their identifiers, the listing of the identifiers of the processes and the retrieval of a special object, called reference, on the basis of the process identifier.

A processer is a runtime service that has the duty of creating new processes in the local runtime node. The creation is performed on the basis of the qualified name of the class implementing the process and a list of initialization parameters.

A dispatcher is a runtime service that allows the exchange of messages among both local and remote processes. In fact, it maps local identifiers to the corresponding process message queue and remote identifiers to the corresponding connection.

The lists of message filters cannot be directly modified by the processes, but they can do it by taking advantage of a filterer. A filterer is a runtime service that allows the creation and modification of the lists of message filters associated with the processes of the local runtime node. Therefore, a process can use such a service for managing the lists of its message filters, and also for modifying the lists of message filters associated with the other processes of the local runtime node.

The distribution model has the goal of defining the software infrastructure that enables the communication of a runtime node with the other nodes of an application possibly through different types of communication supports, guaranteeing a transparent communication among their processes. This model is based on three kinds of element: *distributor*, *connector* and *connection*.

A distributor has the duty of managing the connections with the other runtime nodes of the application. The distributor manages connections that can be built with different kinds of communication technology through the use of different connectors (see Figure 4). Moreover, a pair of runtime nodes can be connected through different connections.

A connector is a connection handler that manages the connections of a runtime node with a specific communication technology allowing the exchange of messages between the processes of the accessible runtime nodes that support such a communication technology.

A connection is a mono-directional communication channel that provides the communication between the processes of two runtime nodes through the use of remote references. In particular, a connection provides a remote lookup service offering the listing of the remote processes and the access to their remote references.

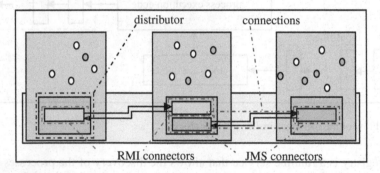

Fig. 4 An application based on three runtime nodes connected through RMI and JMS technologies.

The communication model has the goal of specifying the messages necessary for building the interaction protocols used in an application. In particular, it provides a very simple model for defining the data structures that are shared by the processes of an application and then used as contents of messages. This model is based on two elements: *entity* and *action*. An action is an entity describing a task that can be executed by a process and that might be used by an actor for the execution of such a task to another process.

3.2 Implementation

The HDS software framework has been built taking advantage of the Java programming language. The application architecture model has been defined through the use of Java interfaces and its implementation has been divided in two modules.

The first module contains the software components that define the software infrastructure and that are not directly used by the developer, i.e., all the software components necessary for managing the lifecycle of processes, the local and remote delivery of messages and their filtering. In particular, the remote delivery of messages has been provided through both Java RMI [11] and JMS [10] communication technologies.

The second module contains both the software components that application developers extend, implement or at least use in their code, and the software components that help them in the deployment and execution of their applications. The identification of such software components can be easily done by analyzing what application developers need to implement: *(i)* actor and server classes are used for the implementation of the processes involved in the application; *(ii)* description selector classes are used for the discovery of the processes involved in common tasks; *(iii)* message filter classes are used for customizing the communication among the processes; *(iv)* typed messages are used in the interaction among the processes; and *(v)* artefacts (i.e., Java classes and/or configuration files) are used for the deployment of the runtime nodes, of the communication channels among runtime nodes, and for the start-up of the initial sets of processes and message filters.

Such a second module contains: *(i)* software components for simplifying the implementation of actors, servers, description selectors and message filters (built through four abstract classes called `AbstractActor`, `AbstractServer`, `AbstractSelector` and `AbstractFilter`), *(ii)* a set of messages useful for building typical communication protocols used in distributed applications; and *(iii)* a set of software components (built through an abstract class called `AbstractRunner`) that allow the bootstrap of the runtime nodes of an application and their initialization. With respect to messages and the related communication protocols, the software framework, taking advantage of the communication model, provides the basic actions and entities for building application dependent client-server protocols, for supporting group communication, for building the publish-subscribe protocol, and for implementing FIPA ACL performatives and interaction protocols [5].

4   HDS for the Implementation of Pervasive Information Systems

One of the main reasons why we designed and implemented HDS is to have a software framework that couples flexible and performable distributed software components with the coordination techniques provided by multi-agent systems, for supporting the sharing of information within virtual communities and organizations. Such a need derives from the fact that we have been working on the sharing of information and on collaboration within virtual communities and organizations for about ten years taking advantaged of JADE [2,3], probably the most well known and used multi-agent development framework. The use of such a framework and of agent interoperability standards, i.e., FIPA, allowed us to quickly build very advanced prototypes, but the simple addition of few new features required a lot of time spent in implementing the behaviours of agents, and long validation trials. Moreover, the performance of the developed systems was often poor. Therefore, when a first stable implementation of the HDS software framework became available, we started to work on the implementation a peer-to-peer system that supports the sharing of information and the collaboration among the members of virtual communities by redesigning and extending the last system we built with JADE. Such a system, known as *RAIS (Remote Assistant for Information Sharing)* [13], is a peer-to-peer multi-agent system supporting the sharing of information among a community of users connected through the Internet.

RAIS offers search facility similar to Web search engines, but it avoids the burden of publishing the information on the Web and it guaranties a controlled and dynamic access to information. The use of agents in such a system is very important because it simplifies the implementation of the three main services: *(i)* the filtering of the information coming from different users on the basis of the previous experience of the local user; *(ii)* the pushing of the new information that can be of possible interest for a user; and *(iii)* the delegation of access capabilities on the basis of a network of reputation built by the agents on the community of users. RAIS is composed of a dynamic set of agent platforms connected through the Internet.

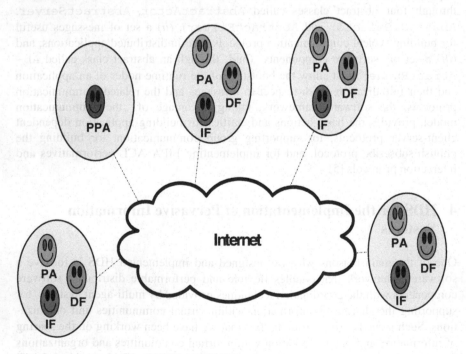

Fig. 5. Architecture of the JADE-based RAIS system.

Each agent platform acts as a peer of the system and it is based on three kinds of agents: a Personal Assistant (PA), an Information Finder (IF) and a Directory Facilitator (DF). Another agent, called Personal Proxy Assistant (PPA), allows a user to access her/his agent platform from a remote system. Figure 5 shows the architecture of the JADE-based RAIS system.

Given the JADE-based architecture of the first implementation of the RAIS system, we designed a new release of the system maintaining a similar architecture, but (i) using a more efficient way to exchange data, which is based on the use of messages via the HDS communication model; and (ii) ensuring system flexibility and evolution through the use of HDS filters. Figure 6 shows the architecture of the HDS-based RAIS system.

In the JADE-based implementation of the RAIS system we spent a lot of work when we tried to add new services. For example, this happened when we added the possibility to exchange documents through the use of encrypted messages and to allow the agent of a platform to exchange information with the agents of other platforms without requiring the authorization of their users on the basis of the reputation of such agents. The management of these two services have been introduced in the new implementation of the RAIS system by adding in each platform a new agent, called Trust and Security Manager (TSM), that monitors the exchange

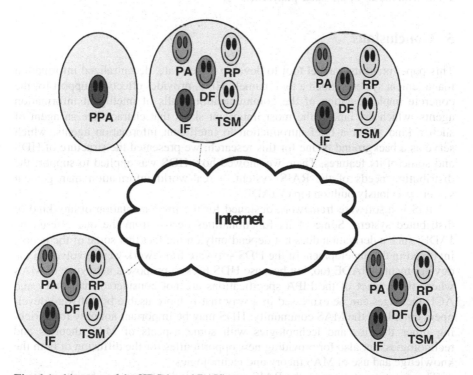

Fig. 6 Architecture of the HDS-based RAIS system.

of messages for building the reputation profiles of the other platforms/users and provides an graphical user interface to its user to view and modify the reputation profiles of the known platform, for defining the access rules for the information she/he like to share with other users and for identifying for which pieces of information an encrypted transmission is required.

The introduction of such a type of agent did not require any modification in the other agents of the system. In fact, a TSM agent: *(i)* acts on the other agents of its platform by adding them filters for modifying the flow of messages from/to the agents of the other platforms; and *(ii)* negotiates with the TSM agents of the other platform when an operation (e.g., the exchange of encrypted messages) requires the introduction of some filters in the agents of the other platforms.

Moreover, in addition to the new TSM agent, we introduced a new agent, called Request Propagator (RP), able to propagate a request received by the IF agent of its platform to the IF agents of some other platforms. Also in this case, the introduction of this type of agent did not require any modification to the other agents of the systems. The duty of an RP agent is to propagate a request if it believes that some other platform might have information for satisfying such a request. To do it, a RP agent builds an information profile for each known platform by adding to its PA agent a filter that captures a copy of each message exchanged by the PA agent with the IF of the other platforms.

5 Conclusions

This paper presents a novel tool to develop large-scale, decentralized information management systems. Such a tool, namely HDS, provides effective support for the concrete implementation of the communication skills of intelligent information agents, which are among the most important skills that characterize an agent of such a kind. After a brief introduction to intelligent information agents, which serve as a background theme for this research, we presented the structure of HDS and some of its features. Then, we showed how HDS was applied to support the distribution needs of the RAIS system, a real-world information management system previously built on top of JADE.

HDS is a software framework designed for the implementation of any kind of distributed system. Some of its functionalities derive from the one offered by JADE; such a derivation does not depend only on the fact that some of the people involved in the development of the HDS software framework were involved in the development of JADE too, but because HDS tries to propose a new view of MAS where the respect of the FIPA specifications are not considered mandatory and ACL messages can be expressed in a way that is more usable by software developers outside of the MAS community. HDS may be important not only for enriching other theories and technologies with some aspects of MAS theories and technologies, but also for providing new opportunities for the diffusion of both the knowledge and use of MAS theory and technologies.

The experimentations on the RAIS system provided interesting results in terms of the effort and the skills needed to implement new features in the systems. In particular, the migration of the system to HDS allowed the fast addition of two kinds of agents with no modifications to the other kinds of agents. This allowed to quickly accomplishing the task of integrating these new kinds of agents and it also required no further testing and validation to the consolidated parts of the system. Unfortunately, such a good result holds when additional agents are only interested in the communication and it may not be so easy when additional agents have functional dependencies upon existing agents. Nonetheless, we believe that the simple APIs of HDS would simplify the development of generic additional agents in comparison to the effort needed to develop them with JADE.

The research presented in this paper has various lines of future development. Current and future research activities on HDS are dedicated, besides to continue the experimentation and validation in the implementation of collaborative services for social network, to the improvement of the HDS software framework itself. In particular, current activities are dedicated to: *(i)* the implementation of more sophisticated adaptation services based on message filters taking advantages of the solutions presented by PICO [9] and by PCOM [1], *(ii)* the automatic creation of the Java classes representing the typed messages from OWL ontologies taking advantage of the O3L software library [12], and *(iii)* the extension of the software framework with a high-performance software library to support the communication between remote processes, i.e., MINA *(http://mina.apache.org)*.

Finally, we intend to take advantage of the experience of porting RAIS to HDS by identifying guidelines and a structured methodology to support developers in similar migrations.

Acknowledgments

This work is partially supported by the Italian Ministry MIUR (Ministero dell'Istruzione, dell'Università e della Ricerca).

References

1. Becker, C., Hante, M., Schiele, G., Rotheemel, K.: PCOM – A component system for pervasive computing. In: Procs. 2nd IEEE Conf. Pervasive Computing and Communications (PerCom 2004), Orlando, FL, pp. 67–76 (2004)
2. Bellifemine, F., Poggi, A., Rimassa, G.: Developing multi agent systems with a FIPAcompliant agent framework. Software - Practice & Experience 31, 103–128 (2001)
3. Bellifemine, F., Caire, G., Poggi, A., Rimassa, G.: JADE: a Software Framework for Developing Multi-Agent Applications. Lessons Learned. Information and Software Technology 50, 10–21 (2008)
4. Bergmans, L., Aksit, M.: Composing crosscutting concerns using composition filters. Communications of the ACM 44(10), 51–57 (2001)
5. FIPA Specifications (2000), http://www.fipa.org
6. Klusch, M.: Intelligent Information Agents. Springer, Berlin (1999)
7. Klusch, M.: Information Agent Technology for the Internet: A survey. Data & Knowledge Engineering 36(3), 337–372 (2001)

8. Klusch, M., Bergamaschi, S., Edwards, P., Petta, P. (eds.): Intelligent Information Agents: The AgentLink Perspective. LNCS (LNAI), vol. 2586. Springer, Heidelberg (2003)
9. Kumar, M., Shirazi, B.A., Das, S.L., Sung, B.Y., Levine, D., Singhal, M.: PICO: A Middleware Framework for Pervasive Computing. IEEE Pervasive Computing 2(3), 72–79 (2003)
10. Monson-Haefel, R., Chappell, D.: Java Message Service. O'Reilly & Associates, Sebastopol (2000)
11. Pitt, E., McNiff, K.: Java.rmi: The Remote Method Invocation Guide. Addison-Wesley, Reading (2001)
12. Poggi, A.: Developing Ontology Based Applications with O3L. WSEAS Trans. on Computers 8(8), 1286–1295 (2009)
13. Poggi, A., Tomaiuolo, M.: Integrating Peer-to-Peer and Multi-Agent Technologies for the Realization of Content Sharing Applications. In: Vargiu, E., Soru, A. (eds.) Information Retrieval and Mining in Distributed Environments, pp. 93–107. Springer, Berlin (2010)

Sensor Mining for User Behavior Profiling in Intelligent Environments

A. Augello, M. Ortolani, G. Lo Re, and S. Gaglio

Abstract. The proposed system exploits sensor mining methodologies to profile user behaviors patterns in an intelligent workplace. The work is based in the assumption that users' habit profiles are implicitly described by sensory data, which explicitly show the consequences of users' actions over the environment state. Sensor data are analyzed in order to infer relationships of interest between environmental variables and the user, detecting in this way behavior profiles. The system is designed for a workplace equipped in the context of Sensor9k, a project carried out at the Department of Computer Science of Palermo University.

1 Introduction

Research in Ambient Intelligence (AmI) focuses specifically on users and on how they relate to the surrounding environment; namely AmI systems attempt to sense the users' state, anticipate their needs and adapt the environment to their preferences [1]. For this reason, AmI systems can benefit from the latest research in users modeling and profiling. User profiling process consists of collecting, and analyzing users informations. The acquired user information can be used to build appropriate user models which can be considered as representation of the system's beliefs about the user [2]. User profiling can be explicit or implicit: an explicit profiling can be done through the formulation of questions about user's preferences, while implicit profiling methods construct user profiles by inferring user ratings from interest indicators by means of user interactions with the system [3]. AmI systems rely on specific hardware in order to gather information about the environment state and the user presence; such data may for instance be collected via pervasively deployed wireless sensor nodes [4], i.e. small devices equipped with sensors, a processor and

Agnese Augello · Marco Ortolani · Giuseppe Lo Re · Salvatore Gaglio
DINFO Dept. of Computer Engineering, University of Palermo, Viale delle Scienze,
ed. 6 – Palermo, Italy
e-mail: {augello,ortolani}@dinfo.unipa.it, {lore,gaglio}@unipa.it

V. Pallotta, A. Soro, and E. Vargiu (Eds.): Advances in DART, SCI 361, pp. 143–158.
springerlink.com   © Springer-Verlag Berlin Heidelberg 2011

a transceiver unit. The idea presented here aims to investigate how such collected data might be profitably used to identify and implicitly profile user habits.

In this context, data mining techniques allow for an intelligent analysis of environmental data in order to detect behavior patterns and classify them into profiles. Careful processing of sensory data may be used to infer descriptive models showing the relationships of interest between environmental variables and the user, while predictive models may proved reliable inference on future behavior of users populating the considered environment [5].

In this work sensor mining methodologies are exploited to profile user behaviors in an intelligent workplace. The workplace has been equipped in the context of Sensor9k, a project carried out at our Department [15]. Our work is based in the assumption that users' habit profiles are implicitly described by sensory data, which explicitly show the consequences of users' actions over the environment state. The system analyzes time data collected by the sensors located in the workplace rooms, and through a data mining process tries to detect changes which can be considered as consequences of user actions. Moreover the sensory data and the recognized events are arranged in appropriate models in order to highlight the existence of relationships among environmental data or events and the users' presence in the office room. The emerging behavioral patterns may finally be grouped based on their relative similarities by means of a clustering process in order to draw users profiles. The rest of the paper is organized as follows: after a discussion about the state of the art, the system architecture will be described, reporting some experimental results and discussing about future works. Moreover the advantages that can be achieved by extending the proposed approach to a distributed architecture will be discussed.

2 Ambient Intelligence for Behavior Profiling

Ambient Intelligence brings intelligence to our everyday environments making those environments sensitive, and adaptive to us [6]. Wireless sensor nodes allow the gathering of data about the environment state [4]. The huge amount of data, obtained from sensor measurements has to be processed and analyzed in order to deduce useful information.

In literature, several methodologies of reasoning on sensor data have been proposed to perform user modeling and profiling, activity recognition and also decision making processes [6]. User profiling applications in AmI have targeted environment personalization as for instance in [8], and [7], where reinforcement learning algorithms are used to learn preferred music and lighting settings, adaptable to preferences changes. User profiling can also be used to detect significant changes in resident's behavior preserving their safety [6][17]. Other applications regard personalization of building energy and comfort management systems [19]. In [9], data collected by wireless sensor are used to create profiles of the inhabitants, and a prediction algorithm allows the automatic setting of system parameters in order to optimize energy consumption. In many projects the aim is to anticipate the location, routes and activities of the residents in order to adaptively control home environments [6]. In MavHome [10], hierarchical models of inhabitant behaviours are learned by means of data-mining techniques aimed to discover periodic and frequent episodes of activity patterns. In the iDorm system [11], intelligent agents are embedded in the user environments to control them according to the needs and preferences of the user. In particular an unsupervised, data-driven, fuzzy technique is used by agents for analyze the actuator readings and sensor states and therefore extracting fuzzy rules representing users behaviors in the environment. A depth examination of AmI methodologies and applications can be found in [6]. Most systems analyze explicit feedback or implicit feedback deriving from the use of actuators, or data related to the presence of the user obtained for example with motion sensors, RFID sensors and so on. The proposed systems can only access information concerning such environmental features as light, temperature and humidity, and has no direct information about the user except for their presence detected by RFID sensors. Therefore, the main feature of the system is that it exploit different methodologies of analysis and reasoning on these data in order to infer the possible user actions on the environment, namely a careful analysis of the data, discarding what can be attributed to noise or to the natural daily trend of the analyzed environmental variables. Our system, similarly to what was proposed by [19] uses probabilistic reasoning to interpret the possible events; moreover, similarly to [9], it extracts an appropriate representation in order to highlight correlations between events. It also uses clustering methodologies to detect similar user behavior patterns.

3 Sensor Data Mining

Data mining methodologies play a fundamental role in sensor data understanding. The main purpose of a data mining process is to find meaningful patterns, relationships and models in a huge amount of data. Pattern discovery methodologies are designed to automatically find new patterns in a predictive way, prefiguring the future behavior of some entities, or in a descriptive way, finding human-interpretable patterns [12]. The various data mining techniques can be classified according to the main categories of applications.

Among predictive methodologies, classification techniques are used to identify the features indicating that an entity belongs to a certain class, learning from a set of pre-classified examples, while prediction and regression techniques analyze known values to estimate unknown quantities. Among descriptive methodologies, clustering techniques are aimed to identify groups with homogeneous elements, association techniques are exploited to identify elements that often appear together during a specific event and sequential patterns are aimed to the detection of recurrent behavior in time sequences of events [13].

Predictive data mining algorithms are used in mining sensory data from an indoor environment to estimate physical at a location where no sensor is placed, to predict failures or user behaviors, while descriptive methodologies can be exploited to find the relationship between variables of interest, as an example to relate users behavior with energy consumption or system failures [5].

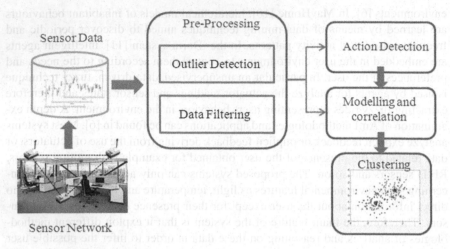

Fig. 1 System Architecture

4 System Architecture

The proposed system aims to learn users' behavior profiles in the context of a smart workplace, such as that of the Sensor9k project [15]. Office rooms have been equipped with sensor nodes monitoring indoor and outdoor physical quantities such as relative humidity, temperature, and light exposure; additionally, RFId sensors allow for detecting the employees' presence in the workplace through the use of personal badges.

The overall system architecture, shown in Figure 1, has been designed according to a modular approach. A *preprocessing* module is used to improve the quality of sensory data while an *action detection* module analyzes data trends to infer changes which can be ascribed to human actions. The information extracted by sensors, and the recognized actions are arranged in appropriate models also keeping into account parameters of interest, such as timing information or any particular environmental conditions. A *correlation* module is devoted to find relationships among the information described in the models; finally a *clustering* module allows to classify the patterns extracted by the correlation module. The following sections will outline the most relevant features of each of the mentioned components.

4.1 Data Preprocessing

The data collected by the sensors are often affected by errors, due to imprecise measurements or to environmental noise. This module is devoted to the detection and removal of invalid values in raw sensory data and to possibly estimating missing data. Initial filtering can be performed assuming that there exist some admissible ranges for the values of the observed variables, and removing all values out of that range. Moreover, spacial and time redundancy can be exploited to detect anomalies in data, or to replace missing data. In particular time redundancy consists in the correlation between consecutive observations read from a sensor, while spatial correlation regards readings from neighboring sensors at a given time [5].

Time correlation can be exploited for the estimation of missing data by means of a linear interpolation between preceding and subsequent observations. Also spatial correlation may be exploited for sensors located in small indoor environments, such as office rooms; in this case, it is reasonable to assume that nearby sensors will retrieve similar measurements, due to the intrinsic nature of the considered physical quantities, so that analysis techniques pointing out significant differences in sensed data may be effectively identify potential outliers. On the other hand, measurements collected in such small indoor areas are typically subject to diverse influencing factors, not necessarily referable to natural phenomena; for instance, the influence of actuators (such as the air conditioning systems, as regards temperature and humidity, or artificial lights, as regards ambient light) cannot be disregarded altogether.

Taking into account the location of sensors within the environment it is thus possible to detect areas in which sensors readings should be correlated at a time point. However, for our aims, it is also important to analyze differences between sensors belonging to different, but close areas. In fact, in this case, variations in sensors readings can be due to the use of actuators from users. So at this stage of preprocessing, spatial correlation property among intra-area sensors can be profitably employed in order to detect outliers, and to replace them with the combination of neighbor sensors readings, while later, the action detection module will analyze in depth the inter-area differences.

4.2 Action Detection

This module aims to perform a deeper analysis of the observed variables trends ,and to recognize those events that can be ascribed to human intervention. We assume here that sudden changes in observed values can be consequence of users' actions, such as turning on/off the light or changing the settings for the temperature and humidity control systems.

In this phase of analysis we consider the placement of the sensors in areas within the environment and analyze the dynamics of the observed series, i.e., the mechanism by which they evolve over time. In particular, the time series obtained from sensor readings can be decomposed in order to detect and remove repetitive and regular changes in data, so as to consider only meaningful changes in the data trend. Finally, a probabilistic inference process can be performed to associate the changes detected in the series to possible user actions on the environment.

4.2.1 Decomposition of Sensory Data Time Series

The analyzed time series obtained from sensor readings can be decomposed into a set of components: a *trend*, a *seasonal* and a *remainder* component [20].

The trend component T_t defines the long-term trend of the variable and can be defined as the tendency to increase, decrease or remain constant over a long period of time; it varies slowly over time and essentially determines the level of the series.

The seasonal or periodic component S_t is given by one or more periodic components, characterized from taking the same or similar values at a fixed distance in time. The remainder component R_t determines short-term fluctuations in the series. The three components can contribute to the composition of the observed series y_t in different ways, additive ($y_t = T_t + S_t + R_t$), multiplicative ($y_t = T_t S_t R_t$), or multiplicative with an additive remainder component ($y_t = T_t S_t + R_t$).

The techniques of decomposition of a time series into its components depend on its composition model. This decomposition is important in order to estimate and remove a regular and predictable component which could hide useful information. In fact, changes in data can be due to regular and repetitive factors, for example to natural light and temperature changes during the day, while other changes can be due to human actions on actuators.

4.2.2   Analysis of Changes in Time Series Data

Let R_t the remainder of the analyzed series, and R_t' the corresponding derivative, representing its time variation. The function R_t' shows changes in R_t function. Every change is characterized by a strength and a direction.

If the detected change is attributed to a user action, the direction will allow the interpretation of the type of action. For example a positive direction will be considered as an turning on of the light, or in the case we are analyzing temperature data could be interpreted as an increase of temperature settings. Moreover the strength allows to analyze a subset of detected changes. In fact, given a threshold experimentally defined, called ϑ, only changes with strength greater than ϑ will be considered.

4.2.3   Probabilistic Inference for Action Detection and Interpretation

Probabilistic models, based on dynamic Bayesian networks, are then used to estimate which of those events may be in fact associated to human actions and to interpreting them. The choice of using these knowledge models depends on the fact that we are bound to reason about uncertain knowledge. We can model the interest variables and their possible states as nodes in the network, connected with directed links representing the influence among the nodes. Influence of parent nodes on a variable X_i is quantified by means of conditional probabilities $P(X_i|ParentsValues)$ represented in opportune tables associated to each node.

In particular, two kinds of variables will be modeled, UA variables, representing possible user actions and SO variables, representing sensory observations. Therefore the built model is used to estimate the probability that the event we want to investigate did occur, based on the sensor readings. As an example, Figure 2 shows the Bayesian network used to estimate the user actions controlling ambient light.

This network represents the possible action on the light settings, by means of the *UserAction t* variable which can take three different states: *on*, corresponding to turning on action, *off*, corresponding to turning off action, and *none* if no action is carried out.

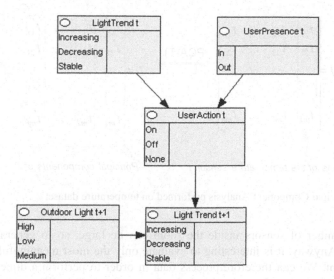

Fig. 2 A Probabilist model to estimate user actions controlling ambient light.

We also have four variables of type *SO*: *UserPresence*,*OutdoorLight t +1*, *Light-Trend t* and *LightTrend t + 1*. The first can assume the states *in* and *out* depending on users presence in the room. Variable *OutdoorLight t +1* represents the external light at a subsequent time instant and can be *high*, *medium*, or *low*; it is used to better understand how much the light change in the room at a subsequent instant may be due to the external light status or to an user action. The variables *LightTrend t* and *LightTrend t + 1* represent the evolution of light in two successive time instants, and can assume the states *increasing*, *decreasing* and *stable*.

The actions are modeled as states of the *UserAction t*, which depends on the state of the light *LightTrend t* variable) and by information on the users presence (*UserPresence t* variable) in workplace. The state of *UserAction t* variable and the state of the outdoor light *OutdoorLight t +1* influence in turns the state of the light at the next istant (*LightTrend t + 1* variable).

4.3 Modeling and Correlation

The correlation module is devoted to identifying relationships among the extracted information. For example it is possible correlate users with sensors measurements within an environment. Matrix models are used to represent the values recorded by different sensors for what concerns a given physical variable during a specific period (such as for instance an entire working day) and to represent the occurrences of events such as user actions in specific instants. Therefore, for each analyzed variable we can build a matrix, whose rows represent the different observations in a given period, while the columns the sample values detected from each sensor during the observations.

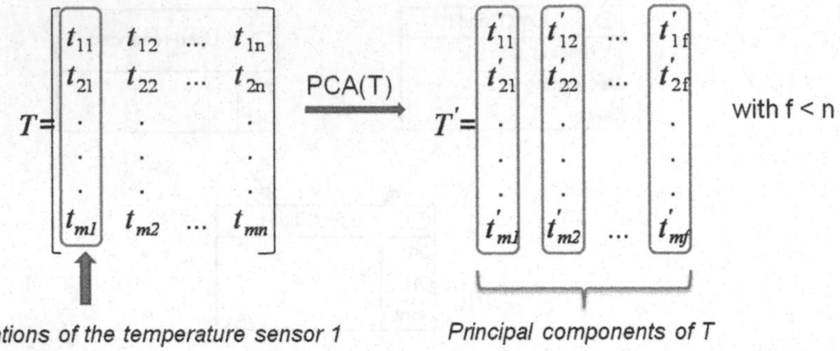

$$T=\begin{bmatrix} t_{11} & t_{12} & \cdots & t_{1n} \\ t_{21} & t_{22} & \cdots & t_{2n} \\ \cdot & \cdot & & \cdot \\ \cdot & \cdot & & \cdot \\ \cdot & \cdot & & \cdot \\ t_{m1} & t_{m2} & \cdots & t_{mn} \end{bmatrix} \xrightarrow{\text{PCA(T)}} T'=\begin{bmatrix} t'_{11} \\ t'_{21} \\ \cdot \\ \cdot \\ \cdot \\ t'_{m1} \end{bmatrix}\begin{bmatrix} t'_{12} \\ t'_{22} \\ \cdot \\ \cdot \\ \cdot \\ t'_{m2} \end{bmatrix} \cdots \begin{bmatrix} t'_{1f} \\ t'_{2f} \\ \cdot \\ \cdot \\ \cdot \\ t'_{mf} \end{bmatrix} \quad \text{with } f < n$$

Observations of the temperature sensor 1   Principal components of T

Fig. 3 Principal Component Analysis performed on temperature dataset

The number of sensors inside the room can be large, so to generate several columns. Anyway, it is interesting to evaluate only the most meaningful informative content. We can therefore process data in order to perform a dimensionality reduction and then evaluate the correlation between variables of interest.

4.3.1 Dimensionality Reduction

The set of observations is to undergo a dimensionality reduction process, by means of PCA [21]. Principal components analysis is a methodology which allows the projection of original dataset space into a smaller space: by a linear transformation of the original variables, PCA extracts a set of orthogonal vectors, called principal components, and arranges them according to decreasing variance values. This transformation has the effect to capture the major associational structure in the dataset, removing information which contribute less to the variance of data, and are thus less relevant. It should be highlighted that PCA performs the dimensionality reduction process, by a combination of original vectors, while other methods merely select a subset of items from the original dataset [22].

Let X indicate the matrix representing a dataset composed by a set of m vectors of length n, each one representing the set of measurements obtained by the n sensor at a specific observation for a specific variable x $X = [x_1, \ldots x_n]$.

As an example, let $T = [T_1, \ldots T_n]$ be a $m \times n$ matrix composed of a set of column vectors, each one representing the set of observations regarding the temperature measured by n sensors in an office room. After performing PCA we obtain a $m \times f$ matrix, with $f \leq n$, $T' = [T'_1, \ldots T'_f]$ (see Figure 3). The same procedure can be performed over the entire set of observed variables.

4.3.2 Correlation among Variables of Interest

The entire set of variables and events observations is then represented as a matrix $X(m \times nv)$, where a row X_i represents an observations at a specific time i and a specific column $\mathbf{X}_j$ represents the entire sample of observations of the j-th variable in the considered period.

In our specific case, matrix $\mathbf{X}$ is given by:

$$\mathbf{X} = \left[\mathbf{U}_1, \ldots \mathbf{U}_d, \mathbf{T}_1, \ldots \mathbf{T}_f, \mathbf{L}_1 \ldots \mathbf{L}_g\right]$$

composed of a set of vectors, each one represents the set of observations of a specific variable. In particular the set $\mathbf{U} = \{\mathbf{U}_j\}_{j=1\ldots d}$ represents observations about the presence of d users in the considered period, while sets $\mathbf{T} = \{\mathbf{T}_j\}_{j=1\ldots f}$ and $L = \{\mathbf{L}_j\}_{j=1\ldots g}$ represent observations about temperature and light exposure respectively, related to the f and g variables obtained after the application of PCA on temperature and light matrices as described in the previous section.

The correlation matrix $\mathbf{C}(nv \times nv)$ is then computed in order to highlight the relationships among the variables. The i, j-th element of $\mathbf{C}$ is given by the correlation coefficient c_{ij} between the i-th and the j-th variable, as given by:

$$c_{ij} = Corr(\mathbf{X}_i, \mathbf{X}_j) = \frac{\sigma_{ij}}{\sigma_i \sigma_j}$$

where σ_{ij} is the covariance between $\mathbf{X}_i$ and $\mathbf{X}_j$ and σ_i and σ_j are respectively the standard deviation of $\mathbf{X}_i$ and $\mathbf{X}_j$.

In this way is possible extract sub-matrices, representing correlation patterns between the observations related to the presence of users in office rooms and values representative of specific environment variables. It is thus possible to obtain a characterization of users with respect to values of the observed variables in a specific period.

4.4 Clustering

The clustering module allows to classify the pattern extracted by the correlation module. The clustering leads to the subdivision of users' behavior patterns into a set of profiles based on their similarities. In this way we can group users with similar preferences about variables setting, or users performing the same actions in similar environment conditions. In particular the K-means [23] algorithm can be used to classify data from the sub-matrices extracted from the correlation matrix $\mathbf{C}$.

The algorithm requires the number of clusters to be obtained and a distance function to evaluate distances between data points and cluster centers. These parameters can be experimentally defined. An iterative process is then performed, during which k data in the dataset are randomly chosen to constitute the first centroids of the clusters. Then the metric distance allows to assign the remaining data to the cluster on the strength of their closeness with the centers of clusters. Then, new centers are detected evaluating the average of each cluster. The process ends when the obtained result satisfies a predetermined criterion of termination.

5   Analysis of Sensor9k Dataset

In this section we report first evaluations of the proposed system on Sensor9k dataset. A set of experiments have been conducted analyzing data measured from MTS300 sensor nodes, where the analyzed variables are light, and temperature. Additional information has been collected to examine users presence, namely the status of the door of the office room, and outdoor light measurements.

In particular we have analyzed light and temperature data regarding two office rooms, *Room1* and *Room2*. The former is an office room, used by two employees *User1* and *User2*, whereas the latter in a common area. The two rooms share similar exposition (thus similar trends for the considered variables), and are connected by a door. We have analyzed data measured from two sensors per room. We will indicate light and temperature measurements collected by the two sensors in *Room1* as *Light101*, *Light102*, *Temperature101* and *Temperature102*, respectively; analogously, *Light201*, *Light202*, *Temperature201* and *Temperature202* will be the measurements related to *Room2*.

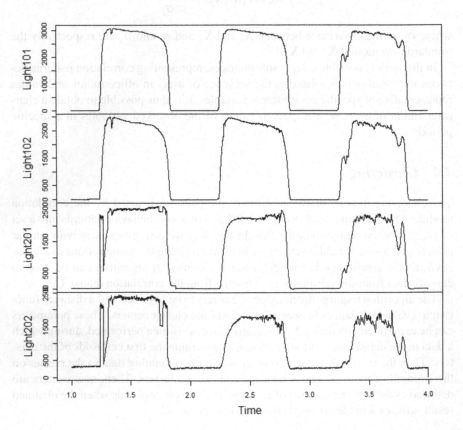

Fig. 4 Time Series for *Light101*, *Light102*, *Light201* and *Light201*

5.1 Analysis of Light Time Series for the Analyzed Rooms

Figure 4 shows the time series of the all four light measurements in a period of three days. The two topmost plots clearly show the similarity in the trends of sensors located within the same room, and the same consideration holds for the two plots at the bottom; if we consider the two central plots, we can also identify significant similarities, although not as striking as in the previous cases; we argue that the differences in measurements for sensors in different rooms are partially due to different placements, and mainly to the effects caused by a different use of the actuators by the users.

Figures 5 and 6 show the decomposition of two light time series belonging to the two rooms, i.e. *Light101* and *Light201*. Each figure shows the original series, and its seasonal, trend and remainder components. The plots show how the seasonal component is related to day-night cycles, while the trend is related to level of brightness of days.

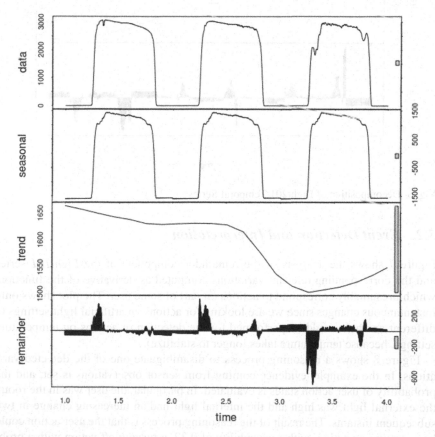

Fig. 5 Decomposition of Light101 Temporal Series

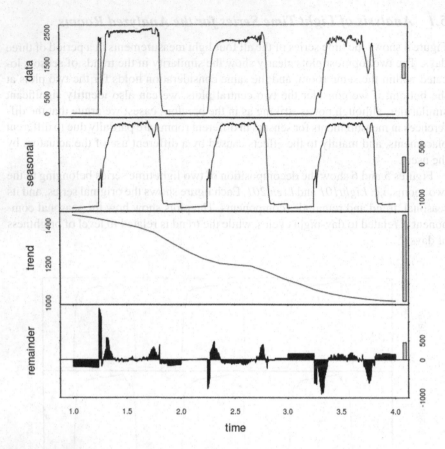

Fig. 6 Decomposition of Light201 Temporal Series

5.2 Event Detection and Interpretation

Figure 7 shows the analysis of the remainder component of the *Light101* series and the corresponding relevant variations computed as derivative of the function, which presumably correspond to actions on part of some user. The plot shows only instantaneous changes since we are looking for actions on artificial light settings (a different analysis would be performed for the detection of actions on temperature settings, because temperature takes longer to stabilize).

Figure 8 shows a reasoning process to disambiguate one of the detected variations. In the example, evidence coming from sensor observations is set, and the probability of user action states is evaluated. In particular, the user was in the room, the external light was high and the internal light had an increasing change in two subsequent instants. The result of the reasoning process is that the user action could be a *turning on* action, with a probability of 0.32, a *turning off* action with a probability of 0.11, and with a probability of 0.57 the increasing will be due to the increasing of the external light (independent of the user action).

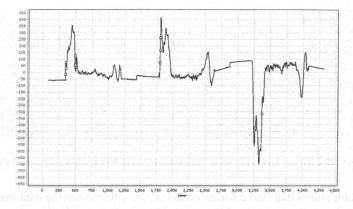

Fig. 7 Light101 Relevant Events: the curve represents the trend of Light, while squares represent the recognized events.

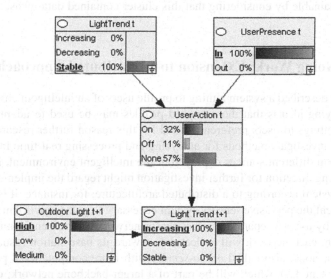

Fig. 8 Probabilist reasoning on a recognized event.

5.3 Clustering of Data

The dataset used to validate our approach so far contains information about the presence of only two users. For this reason, we conducted a proof-of-concept experiment involving a set of time slots in a working day, also considering information about the users' presence and environmental conditions.

In particular, we report here the experiments the experiments related to the clustering module; Figure 9 shows the results obtained by a k-means clustering on one-day matrix observations, related to the presence of the two users *User1* and *User2*, and the measurements of light and temperature.

Attribute	cluster_0	cluster_1	cluster_2	cluster_3
user1	0.307	0.175	0	0
user2	0.631	0.535	0.659	0.148
Light102	2686.862	2402.147	1508.620	9.898
Light101	2603.906	2246.683	1306.187	7.833
Temperature101	25.577	25.729	25.614	9.395
Temperature102	26.273	25.969	26.399	9.609

Fig. 9 Relation between Users presence and Temperatures and Lights values in a day.

The Table shows that *User2* had more significant influence on the measured quantities as it was more present; this is especially evident for the last two clusters, regarding the later time of the day, where the influence of *User2* may be easily singled out; the significant differences in the numerical values captured by cluster 3 are easily explainable by considering that this cluster contained data measured during nighttime.

6   On-Going Work: Extension to a Distributed Approach

This paper described a system aiming to profile users of an intelligent environment; the underlying idea is that the extracted profiles may be used to adapt the environment settings to users preferences, and for this reason further research will be devoted to investigate methods for analyzing and processing real-time information coming from different sources deployed in the intelligent environment. Moreover, an interesting direction for further investigation might regard the implementation of the final system according to a distributed architecture; for instance, it is common to implement the pervasive sensory system necessary to gather environmental measurements by using a separate wireless sensor network for each environment under observation; each network will collect data toward its base station (usually represented by a node, also called *micro-server*, with more computational power than usual sensor nodes), which will be part of a larger backbone network, devoted to ensure connectivity with a remote central storage and processing server. It is thus conceivable to distribute part of the profiling analysis to the local micro-servers, thus relieving the central server from this additional burden; only higher-level information about the profiles will need to be forwarded for further processing. Such architecture is well suited for an agent-based implementation. The analysis of real time data coming from sensors may in fact be performed by a set of autonomous agents, each of which responsible for a particular location. In this way, each agent would perform computations and independently take decisions without overloading the central unit. Moreover agents must be able to exchange information, for example to periodically update users models. In this case, distributed data mining algorithms, would allow agents to individually analyze and process the different informations coming from the multiple sources of information, exchanging and integrating them.

7 Conclusion and Future Work

In this paper we presented a sensor mining system aimed to profiling occupant behaviors. A specific case of study regards Sensor9k, a project carried out at the Department of Computer Science of Palermo University. Environmental variables are monitored by a sensor network, and a set of modules allow to extract useful information regarding user actions and habits. The system is still under evaluation, but we discusses some preliminary results that appear promising with respect to the detection of basic users' habits based on their use of the actuators, which we implicitly infer through the difference in sensory measurements.

The main problem of evaluation is primarily concerned with the need for more data to analyze in order to better understand users habits. A more detailed and extensive dataset will allow to better validate our system. In future work the inferred information about users behaviors and habits will be formalized in an appropriate user model. The creation of this model will have different applications. In particular the users models can be used to adapt the intelligent environment to the user's preferences, or vice versa in order to identify wrong behavior of employees in order to reduce the energy consumption of the building.

References

1. Vasilakos, A., Pedrycz, W.: Ambient Intelligence, Wireless, Networking, Ubiquitous Computing. Artech House Press, MA (2006)
2. Froschl, C.: User Modeling and User Profiling in Adaptive. VDM Verlag (2008)
3. O'Sullivan, D., Smyth, B., Wilson, D.: Explicit vs implicit profiling: a case-study in electronic programme guides. In: Proceedings of the 18th International Joint Conference on Artificial intelligence, Acapulco, Mexico, August 09-15, pp. 1351–1353. Morgan Kaufmann Publishers, San Francisco (2003)
4. Akyildiz, I.F., Su, W., Sankarasubramaniam, Y., Cayirci, E.: A survey on sensor networks. IEEE Communications Magazine 40(8), 102–114 (2002)
5. Wu, S., Clements-Croome, D.: Understanding the indoor environment through mining sensory data–A case study. Energy and Buildings 39(11), 1183–1191 (2007) ISSN 0378-7788, doi:10.1016/j.enbuild.2006.07.011
6. Cook, D.J., Augusto, J.C., Jakkula, V.R.: Ambient intelligence: Technologies, applications, and opportunities. Pervasive and Mobile Computing 5(4), 277–298 (2009) ISSN 1574-1192, doi:10.1016/j.pmcj.2009.04.001
7. Mozer, M.C.: Lessons from an adaptive home. In: Cook, D.J., Das, S.K. (eds.) Smart Environments: Technology, Protocols, and Applications, pp. 273–298. Wiley, Chichester (2004)
8. Khalili, A., Wu, C., Aghajan, H.: Autonomous Learning of User's Preference of Music and Light Services in Smart Home Applications. In: Behavior Monitoring and Interpretation Workshop at German AI Conf. (September 2009)
9. Barbato, Borsani, L., Capone, A., Melzi, S.: Home Energy Saving through a User Profiling System based on Wireless Sensors. In: ACM Buildsys 2009 (in Conjunction with SenSys 2009), Berkeley, CA, November 3 (2009)

10. Youngblood, G.M.: Automating inhabitant interactions in home and workplace environments through data-driven generation of hierarchical partially-observable Markov decision processes. PhD thesis, The University of Texas at Arlington (2005)
11. Doctor, F., Hagras, H., Callaghan, V.: A fuzzy embedded agent-based approach for realizing ambient intelligence in intelligent inhabited environments. IEEE Transactions on Systems, Man, and Cybernetics, Part A 35(1), 55–65 (2005)
12. Fayyad, U., Piatetsky-Shapiro, G., Smyth, P.: From data mining to knowledge discovery in databases. AI Magazine 17, 35–37 (1996)
13. Dunham, M.H.: Data Mining, Introductory and Advanced Topics. Prentice-Hall, Englewood Cliffs (2002)
14. Cantoni, V., Lombardi, L., Lombardi, P.: Challenges for Data Mining in Distributed Sensor Networks. In: International Conference on Pattern Recognition (ICPR 2006), vol. 1, pp. 1000–1007 (2006)
15. De Paola, A., Farruggia, A., Gaglio, S., Re, G.L., Ortolani, M.: Exploiting the Human Factor in a WSN-Based System for Ambient Intelligence. In: CISIS 2009, pp. 748–753 (2009)
16. De Paola, A., Gaglio, S., Re, G.L., Ortolani, M.: Human-ambient interaction through wireless sensor networks. In: Proceedings of the 2nd IEEE Conference on Human System Interactions, pp. 61–64 (2009)
17. Akhlaghinia, M.J., Lotfi, A., Langensiepen, C., Sherkat, N.: Occupant Behaviour Prediction in Ambient Intelligence Computing Environment. Special Issue on Uncertainty-based Technologies for Ambient Intelligence Systems 2(2) (May 2008)
18. Fawcett, T., Provost, F.J.: Combining data mining and machine learning for effective user profiling. In: Proceedings of the Second International Conference on Knowledge Discovery and Data Mining (KDD), pp. 8–13 (1996)
19. Dong, B., Andrew, B.: Sensor-based Occupancy Behavioral Pattern Recognition for Energy and Comfort Management in Intelligent Buildings. In: Proceedings of Building Simulation '2009, an IBPSA Conference, Glasgow, U.K (2009)
20. Brockwell, P.J., Davis, R.A.: Time Series: Theory and Methods. Springer, Heidelberg (1998) ISBN: 038797429
21. Jolliffe, I.T.: Principal Component Analysis, p. 487. Springer, Heidelberg (1986) ISBN 978-0-387-95442-4, doi:10.1007/b98835
22. Han, J., Kamber, M.: Data Mining: Concepts and Techniques, 2nd edn. The Morgan Kaufmann Series in Data Management Systems, Gray, J. Series Editor. Morgan Kaufmann Publishers, San Francisco (2006) ISBN 1-55860-901-6
23. MacQueen, J.B.: Some Methods for classification and Analysis of Multivariate Observations". In: Proceedings of 5th Berkeley Symposium on Mathematical Statistics and Probability, vol. 1, pp. 281–297. University of California Press, Berkeley (1967)

Motivating Serendipitous Encounters in Museum Recommendations

Leo Iaquinta, Marco de Gemmis, Pasquale Lops,
Giovanni Semeraro, and Piero Molino

Abstract. Recommender Systems try to assist users to access complex information spaces regarding their long term needs and preferences. Various recommendation techniques have been investigated and each one has its own strengths and weaknesses. Especially, content-based techniques suffer of overspecialization problem. We propose to inject diversity in the recommendation task by exploiting the content-based user profile to spot potential surprising suggestions. In addition, the actual selection of serendipitous items is motivated by an applicative scenario. Thus, the reference scenario concerns personalized tours in a museum and serendipitous items are introduced by slight diversions on the context-aware tours.

1 Background and Motivation

Recommender Systems (RSs) try to assist users to access complex information spaces. They provide the users with personalized advices based on their needs, preferences and usage patterns.

Various recommendation techniques have been investigated and they have been often evaluated against some accuracy metrics. Nevertheless, the "accuracy does not tell the whole story" [3]. Indeed, the accuracy metrics have their place in retrospective analysis of static datasets but they should cope with user ratings that often are neither precise nor consistent. Moreover, the provided suggestions affect users behavior and the resulting effect is difficult or impossible to model with static datasets. Further, the quest for accuracy may also be misguided, since users do not need precise predictions in order to obtain effective advices in their decision making. Thus, the evaluation of recommender systems should take into account other factors rather than only the predictive accuracy of algorithms [6].

Leo Iaquinta · Marco de Gemmis · Pasquale Lops · Giovanni Semeraro · Piero Molino
Università degli Studi di Bari "Aldo Moro" - Dipartimento di Informatica,
via E. Orabona 4, Bari (Italy)
e-mail: {iaquinta,degemmis,lops,semeraro}@di.uniba.it,
    piero.molino@gmail.com

V. Pallotta, A. Soro, and E. Vargiu (Eds.): Advances in DART, SCI 361, pp. 159–167.
springerlink.com   © Springer-Verlag Berlin Heidelberg 2011

Common expectations for RSs are relevance, novelty and surprise. Neverthe-less, some recommendation techniques, e.g. the content-based ones, suffer of the *over-specialization* problem. Indeed, sometimes RSs can only recommend items that score highly against the user's profile and, consequently, the user is limited to obtain advices only about items too similar to those she already knows. Thus, the user can perceive the recommend items as obvious advices that are not so novel nor surprising. Indeed, novelty occurs when the system suggests an unknown item that the user might have autonomously discovered and a serendipitous recommen-dation helps the user to find a surprisingly interesting item that she might not have otherwise discovered (or it would have been really hard to discover) [7].

According to André et al. [1], the belief of serendipity as a valuable part of cre-ativity, discovery and innovation is the main motivation of the interest of computer scientists about serendipity. Consequently, they have attempted to develop systems that deliberately induce serendipity and celebrated when it appeared as a side ef-fect in systems built with other purposes in mind, for example the serendipitous discovery of something when browsing rather than searching hypertext documents [9]. However, most systems designed to induce or facilitate serendipity focus on the accidental nature of the serendipity and they neglect the breakthrough or discovery made by drawing an unexpected connection. Truly, the connections, though they may be guided, must remain unlooked for specifically to be considered serendipi-tous. Computer systems, however, may be able to help potential discoverers be as primed as possible to make unexpected connections in such a way that they are able to take advantage of them.

Our objective is to try to feed the user also with recommendations that could possibly be serendipitous. Thus, we propose to inject diversity in the recommen-dation task by exploiting the content-based user profile to spot potential surprising suggestions. In addition, the actual selection of serendipitous items is motivated by the real-world situation when a person visits a museum and, while she is walking around, she finds something completely new that she has never expected to find, that is definitely interesting for her. Thus, the applicative scenario pertains to personal-ized tours in a museum and serendipitous items are introduced by slight diversions on the context-aware tours. Indeed, the basic content-based recommender module allows to infer the most interesting items for the active user and, therefore, to ar-range them according the spatial layout, the user behavior and the time constraint. The resulting tour potentially suffers from over-specialization and, consequently, some items can be found not so interesting for the user. Therefore the user starts to divert from the suggested path considering other items along the path with growing attention. On the other hand, also when the recommended items are actually interest-ing for the user, she does not move with blinkers, i.e. she does not stop from seeing artworks along the suggested path. These are accidental opportunities for serendip-itous encounters. The serendipity-inducing module perturbs the optimal path with items that are programmatically supposed to be serendipitous for the active user.

The paper is organized as follows: Section 2 introduces the serendipity issue and covers strategies to provide serendipitous recommendations; Section 3 pro-vides a description of our recommender system and how it discovers potentially serendipitous items in addition to content-based suggested ones; Section 4 provides the description of the experimental session carried out to evaluate the proposed ideas; finally, Section 5 draws conclusions and provides directions for future work.

2 Serendipitous Recommendations

The idea of serendipity has a link with de Bono's "lateral thinking" [4] which consists not to think in a selective and sequential way, but accepting accidental aspects, that seem not to have relevance or simply are not sought for. This kind of behavior helps the awareness of serendipitous events, especially when the user is allowed to explore alternatives to satisfy her curiosity as in the museum scenario.

Moreover, serendipitous encounters depend on personal characteristics, e.g. the open minded attitude, the wide culture and the curiosity [11]. Therefore, the subjective nature of serendipity makes hard its conceptualization, its analysis and its implementation [5]. Anyway, programming for serendipity is feasible [2], for instance, by allowing the users to expand their own knowledge and by preserving the opportunity of serendipitous discoveries.

Toms [13] suggests four strategies to introduce the serendipity: 1) Role of chance or 'blind luck', implemented via a random information node generator; 2) Pasteur principle ("chance favors the prepared mind"), implemented via a user profile; 3) Anomalies and exceptions, partially implemented via poor similarity measures; 4) Reasoning by analogy, whose implementation is currently unknown.

In [8] we propose an architecture for hybridizing a content-based RSs by the "Anomalies and exceptions" approach to provide serendipitous recommendations alongside classical ones. Thus, the basic idea underlying the proposed architecture is to ground the search for potentially "serendipitous" items on the similarity between the item descriptions and the user profile. More specifically, the problem of learning user profiles is managed as a binary *Text Categorization* task, since each document has to be classified as interesting or not with respect to the user preferences. Therefore, the set of categories is restricted to *POS*, that represents the positive class (user-likes), and *NEG* the negative one (user-dislikes). The content-based recommendations come of the matching of the concepts contained in the semantic profile and the concepts contained in the descriptions of items to be recommended. The recommended items are ranked according to the classification score against the *POS* and *NEG* classes. Thus, the list will contain on the top the most similar items to the user profile, i.e. the items high classification score for the class *POS*. On the other hand, the items for which the a-posteriori probability for the class *NEG* is higher, will ranked lower in the list. The items on which the system is more uncertain are the ones for which difference between the two classification scores for *POS* and *NEG* tends to zero. The uncertainty on the classification is used to spot items that are not known by the user, since the system was not able to clearly classify them as relevant or not.

The subjective nature of serendipity exposes to a chance of recommendations lead the users to an unsatisfying or useless result. Hence the users in the future may stop following the recommendations or using the RS as a whole [12]. This drawback can be handled in the museum scenario since the serendipitous encounters can be actuated in a productive way [1].

3   Personalized Museum Tours

RSs traditionally provide a static ordered list of items according to the user assessed interests, but they are not aware about context facets concerning the user interaction with the environment. Besides, if the suggested tour simply consists of the enumeration of ranked items, the path can be too tortuous and with repetitive passages that make the user disoriented, especially under a time constraint. Fig. 1 shows a sample tour consisting of the k most interesting items, where the k value depends on how long should be the personalized tour, e.g., it deals with the overall time constraint and the user behavior. Moreover, different users interact with environment in different manner, e.g. they travel with different speed, they spend different time to admire artworks, they divert from the suggested tour. Consequently, the suggested personalized tour must be dynamically updated and optimized according to contextual information on the user interaction with the environment. The optimization task is performed by a genetic approach with a fitness function that relies on the user-sensitive time constraint, the user behavior (i.e., speed and stay times), the user learned preferences and the item layout.

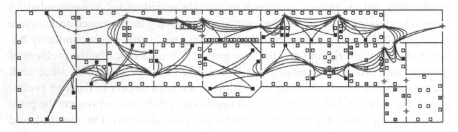

Fig. 1 A sample tour consisting of the ranked k most interesting items

Once the personalized tour is achieved, as shown in Fig. 2, serendipitous disturbs are applied. The diversity injection is pursued by serendipitous disturbs to the personalized tour. Generally, increasing serendipity in a set of recommendations might negatively impact the accuracy that should be perceived as producing of spurious recommendations. Thus, the serendipitous items have to be strategically introduced in order to alleviate the risk of confusing the users or having a wasteful distrust effect. According to Ge et. al [6], this risk can be alleviated by (1) providing explanations about item recommendations and, most importantly, (2) better arranging the recommendation list or using multi-lists. Indeed, a recommendation list usually gives the impression that the recommended items are more or less the same and the top one is the most accurate.

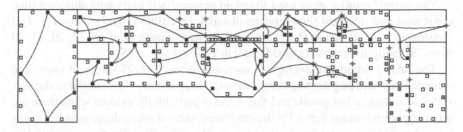

Fig. 2 Optimized version of the tour in Fig. 1

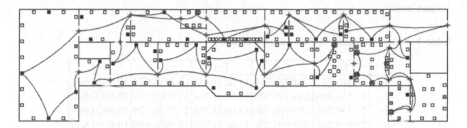

Fig. 3 The "good enough" augmented version

In the museum scenario, the recommendation task produces a personalized tour that is the optimal arrangement of recommended items. Thus, the previous personalized tour is augmented with some items that are along the path and that are in the ranked list of serendipitous items according to the learned user profile and context facets [10]. The resulting path most likely has a worse fitness value and then a further optimization step is performed. However, the further optimization step should cut away exactly the disturbing serendipitous items, since they compete with items that are more similar with the user tastes. Therefore serendipitous items are differently weighed from the fitness function: their supposed stay time is changed. This implementation expedient also deals with the supposed serendipitous items should turn out not so serendipitous and the user should reduce the actual stay time in front of such items. Fig. 3 shows a "good enough" personalized tour consisting of the most interesting items and the most serendipitous ones (circled items in Fig. 3). It is amazing to note that some selected serendipitous items are placed in rooms otherwise unvisited.

4 Experimental Session

The goal of the experimental evaluation is to evaluate the serendipity augmenting effects on personalized tours. The dataset was collected from the official website of the Vatican picture-gallery and it consists of of 45 paintings and 30 users took part in the experiments.

The learned profiles were used to obtain personalized tours with different time constraints and different serendipitous disturbs. Five time (T_{10}, T_{15}, T_{20}, T_{25}, T_{30}) constraints were chosen so that tours consisted approximately of 10, 15, 20, 25, 30 items. Serendipitous items ranged from 0 to 7 (labels $S_0, \ldots, S_7$).

The Table 1 reports the average of sums and means of POS values of tours. The serendipitous item augmenting causes the exploiting of items less similar to the user tastes according to her profile and this effect is particularly evident when there are too many serendipitous items. On the other hand, there is also a decrease when many items are selected according to the user profile, since they are progressively less interesting. When there are many items, the serendipitous item augmenting seems to have no effects over POS mean, but probably this comes from the not very large dataset used.

Table 1 Sums and means of POS values of tours

	T_{10}		T_{15}		T_{20}		T_{25}		T_{30}	
S_0	7.18	0.711	10.69	0.714	14.02	0.705	17.21	0.679	19.94	0.671
S_1	7.15	0.708	10.61	0.709	14.00	0.704	17.20	0.679	19.89	0.670
S_2	7.12	0.705	10.59	0.708	13.98	0.702	17.20	0.679	19.88	0.669
S_3	7.08	0.701	10.60	0.708	13.96	0.702	17.19	0.679	19.87	0.669
S_4	7.03	0.696	10.58	0.707	13.96	0.701	17.19	0.678	19.87	0.669
S_5	6.88	0.681	10.52	0.703	13.95	0.701	17.17	0.678	19.85	0.668
S_6	6.54	0.647	10.42	0.696	13.90	0.698	17.11	0.676	19.75	0.665
S_7	6.17	0.611	10.19	0.681	13.76	0.692	16.99	0.671	19.64	0.661
Items	10.10		14.97		19.90		25.33		29.70	

Table 2 reports percentages of walking time over the tour. Data show that, increasing the time constraint, less time is (relatively) spent to walk. Indeed, if few items are selected, they are scattered around (proportionally) many rooms and the user visits room with very few and even no one suggest item. The serendipitous item augmenting seems to increase the relative walking time. This result is quite amazing according to the selection serendipitous item strategy, i.e., items that are along to a previously optimized path. Actually, the walking time percentage mainly increases because serendipitous items are introduced as new genes of a "good enough" chromosome (solution). However, the augmented chromosome tends to evolve toward the previous one. Thus the new genes should be promoted with a benefit over the fitness function: the reduction in their supposed stay time. This approach is simple and intuitive, but it makes difficult the interpretation of expected walking time percentage. Indeed, the variation on walking time becomes from path variations, but the total tour time is also changed on account of the technical issue about the genetic approach fitness function.

Moreover, the effects of serendipitous items on expected walking time are analyzed with respect to the starting optimized tours (S_0), i.e. the previously discussed drawback is partially cut off. Table 3 shows that few disturbs cause a quite uniform increase of the walking time percentage: the ground becomes from the slight deviations on S_0 tour. On the other hand, growing the number of serendipitous items,

Table 2 Percentages of walking time

	T_{10}	T_{15}	T_{20}	T_{25}	T_{30}	
S_0	39.9	34.0	34.6	31.6	30.2	**34.1**
S_1	42.6	36.3	36.0	32.8	31.3	**35.8**
S_2	45.0	38.1	37.4	34.0	32.2	**37.4**
S_3	49.7	40.1	38.3	34.5	33.5	**39.2**
S_4	52.7	42.0	39.9	36.3	34.6	**41.1**
S_5	56.0	45.5	41.9	37.8	35.9	**43.4**
S_6	60.0	47.5	43.7	39.7	37.2	**45.6**
S_7	65.2	51.7	45.6	41.7	39.0	**48.6**

Table 3 Increment of walking time for tours with serendipitous items

	T_{10}	T_{15}	T_{20}	T_{25}	T_{30}
S_1	106	106	104	103	103
S_2	112	112	108	107	107
S_3	124	119	112	110	111
S_4	131	126	117	115	115
S_5	141	136	123	121	120
S_6	150	143	130	127	125
S_7	164	155	135	134	131

the deviations are amplified. This is more evident for the shortest S_0 tours, since many serendipitous items can encourage the "exploration" of rooms untouched by S_0, about Figure 3.

5 Conclusions and Future Work

This paper presents a beginning effort to apply some ideas about serendipity to information retrieval and information filtering systems, especially in RSs, to mitigate the over-specialization issue. Serendipity has a valuable part of creativity, discovery and innovation, but its subjective nature is problematic when trying to conceptualize, analyze and implement it. The attempts to develop systems that deliberately induce or facilitate serendipity often focus on the accidental nature of serendipity and the delight and surprise of something unexpected. On the other hand, they neglect the breakthrough or discovery made by drawing an unexpected connection. Thus, André et al. [1] stressed the importance of making use of serendipitous encounters in a productive way.

Hence, the evaluation of recommendations has to be further investigated. Indeed, the recommendation process relies on the provided ratings and they should be also interpreted according to the serendipity point of view. Used ratings are often too synthetic and, consequently, they conceal the user rating motivations that affect the meaning evaluation of finding unknown and possibly interesting things, and not simply interesting ones. For instance, a poor rating for suggested items should come from the experience of the user (the user already knows the item), from her lack of interest (the user already knows the item and is not interested in it), from her lack of interest in finding new things (the user does not know the item and has no interest in knowing something new), from the conscious expression of dislike (the user did not know the item before, now she knows it but she does not like it or is not interested in it) or from a serendipitous encounter (before-unknown item that results to be interesting for the user).

The museum scenario is particularly interesting because items are arranged in a physical space and users interact with the environment. Thus disregarding context facets makes useless recommendations.

Similar remarks are still valid in domains (different from cultural heritage fruition) in witch a physical or virtual space is involved and it represents a pragmatic justification to explain (supposed) serendipitous recommendations. Item descriptions are the starting point to exploit content-based methods to implement a hybrid RS that is aware of contextual facets and that uses them, in concurrence with semantic profiles, to spot serendipitous items.

As future work, we expect to carry out more extensive experimentation with more users and wider item collections. We plan also to gather user feedback and feeling by questionnaires focused on qualitative evaluation of the recommendations and the idea of getting suggestions that should surprise them. That is really important for the need to understand the effectiveness of the module in finding unknown items rather the ones that result best rated. Experimentation with users with different cultural levels and with different information seeking tasks are also important to find out which kind of user would like most serendipitous recommendations and to whom they are more useful.

References

1. André, P., Schraefel, m.c., Teevan, J., Dumais, S.T.: Discovery is never by chance: designing for (un)serendipity. In: Proceeding of the 7th ACM Conference on Creativity and Cognition (C&C 2009), pp. 305–314. ACM, New York (2009), doi:10.1145/1640233.1640279
2. Campos, J., de Figueiredo, A.: Searching the unsearchable: Inducing serendipitous insights. In: Weber, R., Gresse, C. (eds.) Proceedings of the Workshop Program at the 4th International Conference on Case-Based Reasoning (ICCBR 2001), pp. 159–164 (2001)
3. Cosley, D., Lawrence, S., Pennock, D.M.: REFEREE: An open framework for practical testing of recommender systems using researchindex. In: Proceedings of 28th International Conference on Very Large Data Bases (VLDB 2002), pp. 35–46. Morgan Kaufmann, San Francisco (2002)
4. De Bono, E.: Lateral Thinking: A Textbook of Creativity. Penguin Books, London (1990)
5. Foster, A., Ford, N.: Serendipity and information seeking: an empirical study. Journal of Documentation 59(3), 321–340 (2003)

6. Ge, M., Delgado-Battenfeld, C., Jannach, D.: Beyond accuracy: evaluating recommender systems by coverage and serendipity. In: Proceedings of the 4th ACM Conference on Recommender Systems (RecSys 2010), pp. 257–260. ACM, New York (2010), doi:10.1145/1864708.1864761

7. Herlocker, J.L., Konstan, J.A., Terveen, L.G., Riedl, J.T.: Evaluating collaborative filtering recommender systems. ACM Transactions on Information Systems 22(1), 5–53 (2004), doi:10.1145/963770.963772

8. Iaquinta, L., de Gemmis, M., Lops, P., Semeraro, G., Molino, P.: Can a recommender system induce serendipitous encounters? In: Kang, K. (ed.) E-Commerce, pp. 227–243. In-Teh (2010)

9. Marchionini, G., Shneiderman, B.: Finding facts vs. browsing knowledge in hypertext systems. Computer 21(1), 70–80 (1988), doi:10.1109/2.222119

10. Mehta, B., Niederée, C., Stewart, A., Degemmis, M., Lops, P., Semeraro, G.: Ontologically-enriched unified user modeling for cross-system personalization. In: Ardissono, L., Brna, P., Mitrović, A. (eds.) UM 2005. LNCS (LNAI), vol. 3538, pp. 119–123. Springer, Heidelberg (2005), doi:10.1007/11527886_16

11. Roberts, R.M.: Serendipity: Accidental Discoveries in Science. John Wiley & Sons, Chichester (1989)

12. Shani, G., Gunawardana, A.: Evaluating recommendation systems. In: Ricci, P.B.F., Rokach, L., Shapira, B., Kantor (eds.) Recommender Systems Handbook, pp. 257–297. Springer, Heidelberg (2011), doi:10.1007/978-0-387-85820-3_8

13. Toms, E.G.: Serendipitous information retrieval. In: DELOS Workshop: Information Seeking, Searching and Querying in Digital Libraries (2000)

Author Index

Angioni, Manuela 1
Armano, Giuliano 13
Augello, Agnese 143

Bergenti, Federico 129

de Gemmis, Marco 159
Delmonte, Rodolfo 59, 81
De Vita, Emanuela 1

Gaglio, Salvatore 143
Ghorbel, Hatem 97

Iaquinta, Leo 159

Jacot, David 97

Lai, Cristian 1, 109
Longo, Laurence 27
Lops, Pasquale 159

Manconi, Andrea 13
Manguin, Jean-Luc 41

Marcialis, Ivan 1
Molino, Piero 159
Moulin, Claude 109

Ortolani, Marco 143

Paddeu, Gavino 1
Pallotta, Vincenzo 81
Poggi, Agostino 129

Re, Giuseppe Lo 143

Semeraro, Giovanni 159

Tiedemann, Jörg 41
Todirascu, Amalia 27
Tripodi, Rocco 59
Tuveri, Franco G. 1

van der Plas, Lonneke 41

Printed in the United States
By Bookmasters